T0273800

Fundamental Mathematical Concepts for Machine Learning in Science

Umberto Michelucci

Fundamental Mathematical Concepts for Machine Learning in Science

Umberto Michelucci
TOELT.AI
Dübendorf, Switzerland

ISBN 978-3-031-56430-7 ISBN 978-3-031-56431-4 (eBook)
https://doi.org/10.1007/978-3-031-56431-4

This Springer imprint is published by the registered company Springer Nature Switzerland AG
The registered company address is: Gewerbestrasse 11, 6330 Cham, Switzerland

To Caterina and Francesca. To Francesca, my love, without whom nothing would be possible. To Caterina, my life, the person I am most proud of in my life.

Preface

In this book, I discuss all the mathematical and methodological aspects that are important when using machine learning techniques in natural sciences (but not only). This book was written with a clear purpose: to explain the complexities of **how to use** machine learning to students and professionals in the natural sciences who may not have a background in computer science. It is a bridge connecting two seemingly disparate worlds and offers a comprehensive guide to understanding and applying the important techniques when applying machine learning in the context of science. I will not discuss algorithms or neural networks, but I will explain all the methods that you need to know to use them properly (e.g. model validation, sampling theory, etc.). The book is structured to gradually build your understanding, starting from fundamental mathematical concepts and progressing to advanced machine learning methods. Each chapter is designed to be self-contained, allowing the reader to focus on specific topics of interest. The chapters on calculus, linear algebra, and statistics are particularly crucial as they lay the foundation for a comprehensive understanding of machine learning algorithms and approaches. Given the breadth and depth of many topics, adequately covering each one would fill many books. My aim has been to cover and explain the core concepts necessary for your journey. I highly recommend further exploring these topics through the additional readings and references that I provide throughout the text. As author, I have striven to present the material in an accessible, yet rigorous manner. The book relies heavily on mathematics as the language of machine learning, ensuring that concepts are conveyed with precision and clarity. Although the book does not delve into programming details, it points out its relevance in machine learning, especially considering Python. What sets this book apart is its focus on methods about **how** to use machine learning, and not on the algorithms themselves filling a gap in the existing machine learning literature. Whether you are a physicist, chemist, biologist, doctor, or involved in any scientific discipline, this book is your guide to harnessing the power of machine learning in your field (if you are a computer scientist, this book is for you too!). This book is a portal to the exciting world of machine learning, written to enrich your scientific research projects, and I hope spark a lifelong interest in this field. If you are an instructor, on the book webpage on Springerlink you will find an instructor guide that will help you to use this book for a course.

Dübendorf, Switzerland

Umberto Michelucci
March, 2024

Acknowledgements

This book would not have been possible without the help of many people. At the forefront is my wife and esteemed colleague, Prof. Dr. Francesca Venturini. Her role in this journey has been immeasurable. Francesca is a constant source of support (in life as in projects), always offering insightful critiques and invaluable suggestions that have significantly shaped the content and direction of this book. I extend my heartfelt gratitude to the numerous students who have patiently listened to my lectures. Their participation and feedback have been instrumental in identifying the most challenging concepts in the field, highlighting areas that are often under-represented in standard university curricula. Their curious minds and challenging questions have not only aided my understanding, but also enriched the content of this book, making it more accessible and relevant to learners at all levels.

A special acknowledgement goes to my editor, Paul Drougas. Paul's expertise, patience, and flexibility have been pivotal in bringing this book to fruition. His guidance through the more challenging phases of this project, combined with his unwavering support, has been invaluable. Paul's proficiency and dedication as an editor have greatly contributed to the quality and coherence of this manuscript.

I am also indebted to many others who have offered their insight, participated in thought-provoking discussions, and provided feedback in various forms. Among them, I am particularly grateful to Dr. Michela Sperti, Prof. Dr. Marco Deriu, and Prof. Dr. Els Ortibus. Each of them has contributed to the richness of the content presented in this book.

Although I have attempted to acknowledge as many contributors as possible, I apologise to those I may have inadvertently omitted. Your contributions have not gone unnoticed and I am deeply grateful for your input. Finally, I assume full responsibility for any errors that may remain in the text. The insights and support provided by all mentioned and those unmentioned have been invaluable, but the final product and any shortcomings it may possess are solely my responsibility. This book is not just a reflection of my efforts, but the result of the collective wisdom and support of many brilliant minds. To each and every one of you, I extend my deepest gratitude.

Dübendorf, February 2024

Contents

Acronyms

Acronyms

AE	Autoencoder
AUC	Area Under the Curve
BFE	Backward Feature Elimination
CART	Classification and Regression Trees
CNN	Convolutional Neural Network
DL	Deep Learning
FFNN	Feed Forward Neural Network
FFS	Forward Feature Selection
GAN	Generative Adversarial Networks
GD	Gradient Descent
GBM	Gradient Boosting Machines
kNN	k-Nearest Neighbour
LLM	Large Language Model
MAE	Mean Absolute Error
ML	Machine Learning
MNIST	Modified National Institute of Standards and Technology
MSE	Mean Square Error
NLP	Natural Language Processing
PCA	Principal Component Analysis
ReLU	Rectified Linear Unit
RFE	Recursive Feature elimintation
RL	Reinforcement Learning
ROC	Receiving Operating Characteristic
t-SNE	t-Distributed Stochastic Neighbour Embeddings
SMOTE	Synthetic Minority Oversampling Techniques
SVM	Support Vector Machine
VAE	Variational Autoencoder

Chapter 1
Introduction

It's supposed to be hard. If it were easy, everyone would do it.
Jimmy Dugan

This preliminary chapter serves as the introduction to the book. I begin by delineating the goal: bridging the gap in machine learning literature for natural science students who may not possess an extensive computer science background. In this chapter, I outline the book's structure, and emphasise that its chapters are crafted to be self-contained, enabling readers to focus on topics of specific interest. I discuss the prerequisites for understanding the content and highlight the importance of a fundamental grasp of mathematics at the undergraduate level. Furthermore, I point out optional and advanced material marked for those who wish to delve deeper. These sections, indicated with a star symbol, are designed for more mathematically inclined readers seeking a comprehensive understanding. Sections marked with a large black square indicate additional content that will help students find relevant advanced concepts. This structure ensures that the book is not just an educational tool but also a starting point for further exploration in the evolving field of machine learning.

1.1 Introduction

This book has its roots in the machine learning lectures and courses that I have taught at various universities for scientists in fields such as astrophysics, biology, and medicine in the past years. Although there are numerous machine learning books, they mainly cater to people coming from computer science, making learning challenging for natural science students. These students often do not have the specific computer science background, and almost always their interests differ from the typical issues explored in computer science. Currently, there is a lack of a comprehensive machine learning text tailored to the natural sciences. This book aims to fill that gap.

The more advanced material, which may not be easy to follow for some students due to their background or experience, is clearly marked and can be skipped in a first reading of the book (see for more details Section 1.7). I worked very hard to explain

U. Michelucci, *Fundamental Mathematical Concepts for Machine Learning in Science*,
https://doi.org/10.1007/978-3-031-56431-4_1

the material in the simplest way possible. However, all the material is being laid out rigorously, and no compromise has been made for the sake of simplicity. Thus, this book relies heavily on mathematics, as this is the language necessary to understand and describe machine learning.

Furthermore, I made a serious effort to ensure that each chapter can stand alone and be understood independently. If you are interested in one specific chapter, you should be able to just study it without having to read all previous chapters.

At URL `https://fmcmls-book.org/` you will find errata, additional information on the book and a form to contact me.

1.2 Choice of Topics

The selection of topics for a book in any field, including machine learning, is inherently subjective. Different experts will have varying opinions on which topics are essential and which are not. However, based on my experience teaching machine learning to students of various levels in various scientific fields, I believe that the topics chosen will be beneficial to the majority of learners. The concepts introduced are the theoretical minimum[1] necessary to use and understand machine learning with students of natural sciences in mind (physicists, chemists, biologists, doctors, etc.) and thus misses some important topics that are only relevant to people doing research in computer science or mathematics (for example, optimisation theory is only scratched here).

To provide the reader with an overview of the most important concepts required to understand and use machine learning, a basic classification of key concepts in machine learning is presented in Tables 1.1 and 1.2. In Table 1.1 the fundamental mathematical and model-agnostic concepts are listed, while in 1.2 the model-specific ones are covered. This book covers all the topics in Table 1.1 but none from Table 1.2. Note that in Table 1.2 only the methods and algorithms that are the most used in machine learning in natural sciences are covered. The list is not exhaustive by **any** means and should only give the student a list of methods that he or she should **at least** learn and study. You may have noticed that reinforcement learning has been omitted. This is quite deliberate, as it represents a fundamentally distinct approach and remains (still) somewhat uncommon in scientific projects.

[1] The eminent Russian physicist Lev Landau created a famously hard entrance examination to evaluate his students. Dubbed "The Theoretical Minimum," this exam encompassed all that he thought fundamental for an aspiring young theoretical physicist. Contrary to what its title might suggest, the test was infamously difficult. During Landau's era, only 43 students managed to successfully pass it. You had to sit through the test in Landau's apartment, with him walking around, looking at what you were writing, and emitting various sounds, typically expressing distaste in your solutions. If you are curious, you should read the excellent paper by Ioffe on his experience in being a Landau's student [1].

Fundamentals	
Mathematics	
Calculus	Limit, derivatives in one and multiple dimensions (including the gradient), optimisation, numerical optimisation, extrema and how to find them, gradient descent.
Linear Algebra	Vectors and matrices, the norm, matrix operations, matrix diagonalisation, eigenvectors and eigenvalues, PCA, variance and covariance matrices.
Statistics	Probability, Random variables, random experiments, distributions, moments of a distributions, Bayes theorem, probability density and mass density functions, central limit theorem.
Model Agnostic Methods	
Model validation and selection	Bias-Variance trade off, Hold-out approach, monte carlo cross validation, k-Fold cross validation, leave-one-out cross validation, model selection, data leakage.
Hyper-parameter Tuning	Black-box optimisation, grid and random search, coarse-to-fine optimisation, search on a logarithmic scale.
Unbalanced Dataset Handling	Over- and under-sampling, synthetic minority oversampling techniques (e.g. SMOTE), metrics for unbalanced datasets.
Feature Selection and Importance	Filter, wrapper and embedded methods, exhaustive feature selection, selection by correlation, forward feature selection, backward feature selection.
Sampling Theory	Research questions and hypotheses, survey populations, non-probability and probability sampling, stratification and clustering, stratified sampling, sampling with and without replacement, bootstrap.
Metrics and Evaluation	Accuracy, MSE, MAE, sensitivity, specificity, precision, F_β score, balanced accuracy, confusion matrix, Receiving Operating Characteristics (ROC) and Area under the curve (AUC).

Table 1.1 A taxonomy of the most important concepts that must be mastered to be able to use machine learning effectively in any research project related to science (and not only). This table covers the fundamentals concepts that are necessary. It is important to say that the list is not exhaustive and contains only the fundamental concepts.

Model Specific Methods	
Supervised Learning	
Classification	Logistic regression, K-Nearest neighbours, support vector machines (SVM), decision trees, random forest, naïve Bayes classifier, Gradient Boosting Machines (XGBoost, LightGBM, etc.).
Regression	Linear regression, ridge regression, lasso regression, polynomial regression, support vector regression, decision trees regressors, adaboost.
Unsupervised Learning	
Clustering	K-Means, hierarchical clustering, density-based clustering of applications with noise (DBSCAN), mean shift clustering.
Dimensionality Reduction	Principal component analysis (PCA), t-Distributed stochastic neighbour embeddings (t-SNE), linear discriminant analysis (LDA), autoencoders (deep learning).
Semi-supervised Learning	
Classification	Self-training approach, semi-supervised support vector machines, label propagation and label spreading.
Regression	Co-training algorithms, self-training approach, graph-based methods.
Deep Learning	
Model Training	Back propagation, auto differentiation, transfer learning.
Optimisers	Gradient descent, adam, RMSProp, momentum, adagrad (adaptive gradient algorithm).
Regularisation	ℓ_1, ℓ_2, dropout.
Neural Network Architectures	FFFN, CNN, RNN, Transformers, U-Net, Generative Adversarial Networks, autoencoders.
Ensemble Methods	
Bagging and Boosting	Random forest, adaboost, bagging, extremely randomised trees.
Stacking and voting	Stacked generalisation, voting classifier.

Table 1.2 A taxonomy of the most important concepts that must be mastered to be able to use machine learning effectively in any research project related to science. This table covers the model specific methods and algorithms. The list is **not** exhaustive but covers the most used and known algorithms and methods that an expert in machine learning should know about.

1.3 Prerequisites

To grasp the content of this book, an undergraduate-level understanding of mathematics is sufficient. This refers to the level typically achieved in a bachelor's degree in a scientific discipline. The book includes chapters on linear algebra, calculus, and statistics, providing a basic grasp of these crucial concepts. However, these chapters are not intended to replace a formal undergraduate course. While subjects like linear algebra and calculus are extensively covered in fields such as physics and mathematics, they are often minimally addressed in disciplines like biology or medicine. However, they are vital for understanding many machine learning concepts. For instance, Principal Component Analysis (PCA), widely used in medicine and other fields, is deeply rooted in linear algebra. Without understanding linear algebra, one cannot fully comprehend PCA. Similarly, training neural networks is fundamentally based on calculus; lack of knowledge of derivatives makes it difficult to understand how neural network weights are optimised. These are just a couple of examples among many. A solid understanding of mathematics is crucial for a meaningful grasp of machine learning.

No programming is discussed explicitly in this book, although I try to point out when something is relevant for aspects of machine learning related to computing. The few examples and notes were written with Python in mind.

1.4 Book Structure

The book starts with a chapter on the history of machine learning and discuss some terminology, to make sure that we will speak the same language. In Chapter 3, calculus and optimisation are introduced and discussed, including a long discussion about gradient descent and its variations. Chapter 4 introduces the main concepts of linear algebra as vectors and matrices. Statistics and probability follow in Chapter 5. The chapter includes key topics such as Bayes' Theorem and distribution functions, giving readers the necessary statistical understanding to understand the inner workings of machine learning. Chapter 6 shifts focus to the practical aspects of machine learning, specifically the creation of datasets. It emphasises the importance of proper sampling techniques when creating a dataset to ensure the development of unbiased and representative datasets, a critical step in the machine learning process. Model validation, an essential aspect of machine learning, is explored in the seventh chapter. This section delves into the bias-variance trade-off and various validation techniques, providing insight into how to evaluate and refine machine learning models effectively. The book also addresses the challenges of working with unbalanced datasets and the importance of choosing appropriate metrics for model evaluation in Chapter 8. This discussion is particularly relevant for practical applications of machine learning, where data often present imbalances. Finally, the book concludes with a focus on hyper-parameter tuning in Chapter 9 and model agnostic feature importance determination in Chapter 10. These chapters provide valuable insights into optimising machine learning models and understanding the impact of different features on model predictions.

1.5 About This Book

Machine learning is a vast field with various approaches, making it challenging for one person to master every aspect and theory underlying its inner workings. However, some mathematical principles are consistently relevant across different approaches. Fundamental concepts such as optimisation, matrix operations, model validation, and hyper-parameter tuning are critical, not only for specific algorithms like neural networks, but for all machine learning models. This book explores these essential elements, which are key for practical implementation and understanding of machine learning in real-world research settings. Instead of focusing on particular machine learning areas like natural language processing or computer vision, this text emphasises the mathematical underpinnings necessary to effectively employ machine learning approaches in these specialised fields. It addresses the foundational aspects of the intricate structure of machine learning pipelines. By grasping the topics covered in this book, you will gain the confidence to navigate and work proficiently within the realm of machine learning.

1.6 Warnings, Info and Examples

In the book, my aim is to emphasise key concepts, practical advice, and cautionary notes wherever possible. To facilitate this, I employed the use of coloured boxes, as illustrated in the following examples, to ensure that certain points stand out more prominently.

Warning **Warnings**

Warning boxes contain material that you should pay attention to and that may be tricky to use or understand. For example you will find a discussion about the ReLU function (more on that in the next chapter) and the fact that it does not have a derivative at $x = 0$. This fact, if not handled correctly, may give you issues in your code and thus can be found in a warning box.

Info **Information**

Information boxes contains material that adds material, proofs or simply interesting tidbits to the text. For example you will find the proof that

$\lim_{x \to 0}(\sin x/x) = 1$ that, although interesting, will not impact your understanding of the topic if skipped.

1.7 ★ Optional and Advanced Material

Some of the material in this book is of a slightly more advanced nature and can be skipped in a first reading of the book. Sections that are optional are marked with the star symbol ★ at the beginning of the title (see the title of this section as example). You will also find the optional sections marked with the star symbol in the table of contents.

We will discuss several important theorems in this book, but students may skip the most difficult proofs if they are not so mathematically inclined. Proofs or discussions that are more complex are sometimes inserted in boxes. The title of the box will also have a star symbol at the beginning to remind you that it is more difficult to understand. Definitions may also have a star in the title. For example, a definition may look like this.

Definition 1.1 (★ Definition of something) Some definition...

A section may look like the one you are reading now (see the star at the beginning of the title of this section?). If you encounter a large star symbol throughout the text, it signifies content of a somewhat more challenging nature. This material, while valuable for those with a stronger mathematical background, can be optionally bypassed without hindering the overall comprehension.

1.8 ■ Further Exploration and Reading

In the book, you will find sections marked with a black square (as in the title of the heading above). These sections, rather than offering detailed explanations about some specific concept, present a curated list of advanced topics, each accompanied by key references when applicable. This format allows readers to quickly identify topics of interest and access relevant literature for in-depth study. Covering a wide range of topics from cutting-edge algorithms to niche applications, these sections serve as an efficient springboard for further exploration and self-directed learning.

1.9 References

Each chapter has its own list of references. The list is, when applicable, very comprehensive, and you will find a large amount of material that you can check if you are interested in learning more or in looking at original papers about specific concepts or ideas. If you are looking for a specific reference, your best bet is to insert the title into Google Scholar (`https://scholar.google.com/`). You should find a downloadable version of it easily enough.

1.10 Let us Start

I wish you a lot of fun with this book and I hope you learn how machine learning works and how to use it in your own projects.

References

1. B. L. Ioffe. Landau's Theoretical Minimum, Landau's Seminar, ITEP in the Beginning of the 1950's, April 2002. arXiv:hep-ph/0204295.

Chapter 2
Machine Learning: History and Terminology

The limit of my language means the limit of my world.
Ludwig Wittgenstein

This chapter traces the evolution of machine learning (ML), from its early theoretical foundations to its contemporary applications in various scientific disciplines. Starting with Alan Turing's seminal work and the development of the perceptron in the mid-20th century, We explore briefly key milestones like the introduction of neural networks, the impact of Minsky and Papert's book, and the resurgence of ML in the late 20th century with advancements in algorithms and computational power. The chapter highlights the transition from theoretical research to practical applications, marked by significant developments such as backpropagation, decision trees, support vector machines, and reinforcement learning. The role of ML in fields such as chemistry, physics, and biology is discussed, emphasising its transformative impact on drug discovery, high-energy physics, astrophysics, genomics, and proteomics. The chapter concludes by categorising ML into various types: supervised, unsupervised, semi-supervised, and reinforcement learning, each of which plays a distinct role in advancing scientific knowledge.

2.1 Brief History of Machine Learning

Machine learning appeared for the first time in the mid-20th century. Alan Turing, one of the pioneers, raised fundamental questions about the potential of machines to mimic aspects of human intelligence in his seminal paper "Computing Machinery and Intelligence" [1]. In this work, Turing endeavored to answer the question "can machines think?" and introduces the concept now known as the Turing test. This is designed to evaluate a machine's ability to exhibit intelligent behaviour that is indistinguishable from that of a human. It involves a human interrogator engaging in a conversation with both a human and a machine, both out of sight. The machine is considered to have passed the test if the interrogator cannot consistently tell it apart from the human. Throughout the paper, Turing addresses and counters various objections to the notion of machine intelligence. These objections range from theological arguments to the belief that consciousness is a prerequisite for intelligence,

U. Michelucci, *Fundamental Mathematical Concepts for Machine Learning in Science*,
https://doi.org/10.1007/978-3-031-56431-4_2

and even the idea that machines are incapable of errors or learning from experiences. Turing also discusses the potential of digital computers, which at the time were in their early stages of development. He highlights their ability to simulate any process of formal reasoning, acknowledging their existing limitations but predicting significant advancements in their capabilities. An intriguing part of Turing's paper is his speculation on the possibility of machines learning to evolve over time. He suggests that instead of equipping a machine with an extensive understanding of the world, it might be more effective to develop a simpler machine and allow it to learn from its interactions and experiences (that is, after all, what machine learning is about). This paper is important for its philosophical approach, as Turing ponders over the broader implications and future possibilities of machine intelligence. His insights and hypotheses have profoundly influenced the development of artificial intelligence and continue to shape discussions about the nature of intelligence and consciousness today.

In the late 1950s, the perceptron, the first neural network, was introduced by Rosenblatt [2]. This development laid the groundwork for future research in neural networks. The perceptron model was conceptualised as a simplified mathematical abstraction of a biological neuron and was designed to perform certain types of classification. It was able to automatically learn the optimal weight coefficients, which are then multiplied with input features to determine whether a neuron fires or not, essentially making a decision based on a linear combination of its input signals. Rosenblatt's work was groundbreaking because it demonstrated how a machine could be programmed to learn from data, a concept that was relatively novel at the time. The perceptron was shown to have the ability to learn through a process of adjusting the weights based on errors made in previous predictions, an early form of what would later be known as supervised learning (see next section for an explanation of the term *supervised*). While the original perceptron had limitations, notably its inability to process data that are not linearly separable (which was later addressed by the invention of multilayer perceptrons), Rosenblatt's paper laid the first foundation for further research into neural networks. It sparked a wave of interest and optimism about the potential of machines to learn and make decisions, paving the way for the modern field of machine learning and artificial intelligence.

The book "Perceptrons" by Marvin Minsky and Seymour Papert [3], published in 1969, stands as a critical and influential work. This book is particularly notable for its analysis and criticism of the perceptron by Rosenblatt. Minsky and Papert's work provided a thorough mathematical critique of the perceptron, delving into its capabilities, and more importantly, its limitations. The authors demonstrated that perceptrons, as they were then conceived, were incapable of solving some relatively simple but fundamental problems, such as the XOR problem, which involves correctly classifying inputs that are not linearly separable. This limitation significantly restricted the range of problems to which the perceptron could be applied.

The criticism contained in "Perceptrons" was so compelling that it led to a significant reduction in interest and funding for neural network research. This period, often referred to as the first "AI Winter," was characterised by skepticism and reduced expectations for the field of artificial intelligence, particularly in the area of neural networks. However, in the long run, the rigour and depth of Minsky and Papert's analysis also laid the groundwork for later advancements. Their work highlighted the need for more complex and layered neural network architectures, leading eventually to the development of multilayer perceptrons and deep learning techniques. In this way, "Perceptrons" played a paradoxical role in the history of machine learning: while it temporarily dampened enthusiasm for neural network research, it also set the stage for some of the field's most significant breakthroughs in subsequent decades. The impact of "Perceptrons" is a testament to the importance of critical analysis and rigorous evaluation in the advancement of scientific fields, serving as both a cautionary tale about the risks of hype and untested assumptions, and a beacon pointing the way to more robust and capable models in machine learning.

The renaissance in neural networks was sparked in the 1980s with many important new results. The decade of 1980 to 1990 was a period of significant advancement in the field of machine learning. One of the major developments was the popularisation of decision tree algorithms (first born in the 1960s) for classification tasks. The introduction of the ID3 algorithm by Quinlan in 1986 [4] marked a key moment in this area. ID3 was a revolutionary step using the concept of information entropy, a measure from information theory, to select the attribute that partitions the data in the most informative way at each branch of the tree. Following the success of ID3, further developments and iterations led to more sophisticated algorithms such as C4.5 and CART (Classification and Regression Trees), which expanded and refined the decision-making capabilities of these models. These advances included handling categorical and numerical data, dealing with missing data, and improving computational efficiency. The development of decision trees significantly impacted the field of machine learning, providing a foundation for more complex models and algorithms. Their ability to break down a complex decision-making process into a simpler form not only made them powerful tools in predictive modelling, but also contributed to the field's broader understanding of data-driven decision-making processes.

One of the major advances in the field of neural networks was the development of the backpropagation algorithm by Rumelhart, Hinton and Williams in 1986 [5], which revolutionised neural network training. Backpropagation, formally known as the *backpropagation of errors*, is a method used for artificial neural networks to calculate the weight updates contributed by each neuron after processing a batch of data. It is a cornerstone of neural network training, enabling efficient computation of gradients. Backpropagation's roots can be traced back to the early work in the 1960s and 1970s. However, it was not until the 1980s that the algorithm gained substantial recognition. The paper by Rumelhart *et al.* was crucial because it addressed a significant challenge of the time: how to adjust the weights of neurons in hidden layers of a neural network. Before this, while single-layer neural networks were used effectively, the addition of hidden layers, which allowed the modelling of more complex functions, posed a challenge in terms of adjusting the weights through learning. The backpropagation algorithm uses the chain rule from calculus to iteratively calculate gradients for each layer in the network, starting from the output layer and working backward. This method allowed for the effective training of deep neural networks by adjusting the weights in a way that minimises the loss function of the network. The introduction of backpropagation sparked renewed interest in neural networks, leading to what is often referred to as the second wave of neural network research. The algorithm's ability to train deep networks laid the groundwork for the development of deep learning, which has since revolutionised many aspects of machine learning and artificial intelligence. Despite its success, backpropagation is not without limitations, such as the vanishing gradient problem, where gradients can become increasingly small as they are propagated back to early layers, hindering the learning of the network. Nevertheless, its development represents a pivotal moment in the history of machine learning, providing a key tool for the training of complex neural networks and enabling advances in a wide range of applications, from image and speech recognition to natural language processing.

This period also saw the emergence of the first practical applications of neural networks, such as the NetTalk system by Sejnowski and Rosenberg in 1987 [6], which demonstrated the potential of neural networks in speech synthesis. NetTalk was designed to convert written English text into phonetic pronunciations, essentially teaching a computer how to speak. At its core, NetTalk used a seven-layer neural network, a significant design for its time, considering the computational limitations and the relatively new state of neural network research. This neural network was trained using the backpropagation algorithm, which allows it to learn the appropriate phonetic pronunciation of English text by being exposed to examples of text and the corresponding spoken words. One of the remarkable aspects of NetTalk was its ability to learn and generalise from the training data. As it was exposed to more examples, the neural network improved its pronunciation skills, much like how a human child learns to speak by listening to adults. This learning process was not just about memorising specific pronunciations, but rather about understanding the underlying patterns and rules of English phonetics. The NetTalk system was instrumental in demonstrating practical applications of neural networks, particularly in the domain of speech synthesis. It showed that neural networks could learn to perform tasks without explicit programming of rules, a significant departure from the conventional approaches used in computer science at the time. Furthermore, NetTalk contributed to the growing interest in neural networks and their potential applications. It was an early example that helped pave the way for more advanced speech recognition and synthesis systems, which are now commonplace in devices and applications such as smartphones, virtual assistants, and language translation services. The development of NetTalk by Sejnowski and Rosenberg is a notable example of how innovative applications of neural networks can lead to breakthroughs in artificial intelligence, demonstrating the power of these systems to learn and adapt in ways similar to human learning.

The 1990s was also a pivotal era in the field of machine learning, marked by a transition from theoretical underpinnings to more practical applications. This period witnessed a significant evolution in algorithms driven by increasing computational power and data availability. One of the major milestones was the advancement of neural networks. The work of LeCun *et al.* [7] on handwriting recognition using convolutional neural networks laid the foundation for modern deep learning techniques.

LeCun assembled a dataset that is now widely known as the MNIST (Modified National Institute of Standards and Technology) dataset. This has played a pivotal role in the field of machine learning, particularly in the development and benchmarking of algorithms for handwritten digit recognition. The dataset was created as a more accessible and pre-processed version of the earlier NIST dataset. It consists of 70,000 labelled images of handwritten digits (0 through 9), divided into a training set of 60,000 examples and a test set of 10,000 examples. Each image is a 28x28 pixel grayscale representation of a digit, making it ideal for testing image processing systems. The simplicity and size of the MNIST dataset have made it a standard benchmark for algorithms in image recognition, serving as a testbed for a wide range of approaches from classical machine learning to deep learning. In particular, the development and success of convolutional neural networks were significantly bolstered by experiments on the MNIST dataset, as demonstrated by LeCun *et al.* [8]. Despite its simplicity, MNIST continues to be a valuable resource for educational purposes and initial algorithm testing.

This period also saw the rise of Support Vector Machines (SVMs), introduced in their modern form by Cores and Vapnik in 1995 [9], which provided a robust method for classification and regression tasks. SVMs are a set of supervised learning methods that are used for classification, regression, and outlier detection. The origins of SVMs can be traced back to the work of Vapnik *et al.* from 1964 [10], where the concept of a linear classifier with maximum margin was first introduced. This early model laid the groundwork for what would eventually become modern SVMs. The formal development of SVMs began in the 1990s. The key paper by Boser *et al.* [11], introduced a way to create nonlinear classifiers by applying the kernel trick to maximum-margin hyperplanes. This was further refined and popularised by Cores and Vapnik [9], who provided a comprehensive framework for training SVMs. Since then, SVMs have seen various enhancements and have been applied in numerous fields, from image recognition to bioinformatics. Their ability to handle large feature spaces and their flexibility in modelling diverse sources of data make them a powerful tool in machine learning. SVMs represent a significant development in the field of statistical learning, offering a robust approach to both classification and regression problems. Their evolution from a theoretical concept to a staple in machine learning toolkits is a testament to their versatility and effectiveness.

The 21st century has been particularly marked by the rise of deep learning, a subset of machine learning based on neural networks with a large number of layers and parameters. Pioneering work by researchers such as Yann LeCun, Geoffrey Hinton, and Yoshua Bengio [12] has led to breakthroughs in fields such as image and speech recognition and natural language processing. But the early 2000s also marked a period of significant advancements and milestones in many fields of machine learning, setting the stage for the rapid development that would follow in subsequent years. One of the most impactful changes at the beginning of the 2000s was the dramatic increase in computational power, largely due to the advancement of GPUs (Graphic Processing Units). This, coupled with the growing availability of large datasets, allowed researchers to train more complex models, particularly deep neural networks. The early 2000s also witnessed significant progress in natural language processing (NLP). The introduction of statistical methods over traditional rule-based approaches marked a paradigm shift in NLP. Seminal models like latent Dirichlet allocation (LDA) by Blei *et al.* [13] began to emerge, changing the way machines processed and understood human language. This period also saw the rise of ensemble methods in machine learning. Techniques such as Random Forests and boosting methods became popular due to their robustness and effectiveness in various applications. Boosting, a powerful ensemble technique, involves combining multiple weak learners to form a strong classifier. A pivotal development in boosting was the AdaBoost algorithm, introduced by Freund and Schapire [14]. AdaBoost, short for Adaptive Boosting, works by sequentially adding weak learners, typically decision trees, and focusing on the instances that were incorrectly predicted in previous rounds. The algorithm assigns higher weights to these challenging instances, ensuring that subsequent learners focus more on them. AdaBoost demonstrated remarkable effectiveness in improving the accuracy of classification models. Another critical advance was the formulation of Gradient Boosting Machines (GBMs) by Friedman [15]. GBMs extend the boosting framework by optimising arbitrary loss functions. This method involves building a model in a stage-wise fashion and generalising them by allowing optimisation of an arbitrary differentiable loss function. GBMs have been highly successful in a wide range of practical applications, from standard regression tasks to complex learning problems.

Parallel to these developments, reinforcement learning has evolved, with significant contributions by Sutton and Barto [16]. This area of machine learning, focusing on how agents should take actions in an environment to maximise some kind of reward, has led to impressive displays such as AlphaGo's victories in the game of Go [17]. A foundational concept in reinforcement learning is temporal difference (TD) learning, introduced earlier by Sutton [18]. In the early 2000s, the refinement and application of TD learning methods, particularly TD-Gammon by Tesauro [19], demonstrated the potential of RL in complex domains. TD-Gammon, a computer backgammon program, used a neural network trained through TD learning and was able to achieve a performance level comparable to that of human experts. The beginning of the 2000s also witnessed significant advancements in policy gradient methods, an approach in RL where the policy, the core decision-making function of an agent, is directly optimised. The work by Sutton *et al.* [20] provided a comprehensive framework for these methods, which became crucial for solving tasks that require complex action sequences. The twentieth century also saw important theoretical advances in RL. The development of algorithms with guaranteed convergence and the exploration of the balance between exploration and exploitation were key focus areas. The formalisation of the exploration-exploitation trade-off, as discussed by Auer *et al.* [21], provided a deeper understanding of the decision-making process in RL. The development of machine learning algorithms capable of playing the ancient board game Go marks a significant milestone in the history of artificial intelligence (AI) and machine learning. Go, known for its deep strategic complexity, has long been a formidable challenge for AI researchers. Traditional AI approaches that had succeeded in chess, such as brute-force search, were ineffective for Go due to the game's vast number of possible positions. The breakthrough came with AlphaGo in 2016, developed by Google DeepMind. In the landmark paper by Silver *et al.* [17], the AlphaGo algorithm was introduced. It combined deep neural networks with Monte Carlo Tree Search (MCTS) to evaluate board positions and select moves. This approach allowed AlphaGo to defeat a professional human Go player in 2015, a feat previously thought to be at least a decade away. Following this, AlphaGo Zero, as detailed by Silver *et al.* [22], represented a further advancement. It learnt to play Go solely by playing games against itself, without any human data, achieving superhuman performance. This was a significant step in unsupervised learning within machine learning. The success of AlphaGo and its successors has had a profound impact on the field of machine learning. It not only demonstrated the potential of deep learning and reinforcement learning in solving complex problems, but also inspired numerous applications of similar techniques in various domains. The development of AlphaGo and its successors marks a pivotal point in the history of machine learning, showcasing the power of integrating deep neural networks with advanced search strategies. This milestone in AI research has broadened the horizons for the application of machine learning in complex decision-making scenarios.

The decade 2010 to 2020 was marked by remarkable breakthroughs in the field of machine learning. One of the most significant milestones was the dominance of deep learning. The pivotal moment occurred in 2012 when [23] demonstrated the power of deep convolutional neural networks (CNNs) by winning the ImageNet competition. This breakthrough drastically improved the performance of image classification tasks and sparked a wave of research and applications in deep learning across various domains. In natural language processing (NLP), the development of transformer models, particularly the introduction of BERT (Bidirectional Encoder Representations from Transformers) by [24], marked a significant advance. This model and its architecture revolutionised the approach to tasks such as text classification, sentiment analysis, and question-response, setting new standards for performance in NLP. Another notable development was in generative models, particularly with the introduction of Generative Pretrained Transformer (GPT) models by OpenAI. As described by [25], these models showed an unprecedented ability in generating coherent and contextually relevant text.

The early 21st century saw a shift towards more complex models with the introduction of neural networks in language tasks. A milestone was the development of sequence-to-sequence learning models [26], which improved the performance of machine translation systems. A pivotal moment in the history of large language models (LLMs) was the introduction of the Transformer model by Vaswani et al. in 2017 [27], which led to a paradigm shift in the way language models were designed. This model, based on self-attention mechanisms, significantly improved the efficiency and effectiveness of language understanding and generation. Transformers have revolutionised the field of natural language processing (NLP) and beyond, offering a new paradigm for handling sequential data. Unlike previous models that processed sequences step by step, transformers process entire sequences simultaneously, leading to significant gains in efficiency and effectiveness. The core innovation of transformers is the attention mechanism. This mechanism allows the model to focus on different parts of the input sequence when predicting each part of the output sequence.

In simple terms, the attention mechanism lets the model dynamically pay attention to the most relevant parts of the input to make predictions, akin to how humans focus on specific words or phrases when comprehending a sentence. A transformer model consists of an encoder and a decoder. The encoder reads the input sequence and generates a high-dimensional representation, which the decoder then uses to generate the output sequence. Both the encoder and decoder are composed of layers that include attention mechanisms and feedforward neural networks. The encoder's job is to process the input data (like a sentence in NLP tasks) and create a context-rich representation. Each layer in the encoder consists of two main parts: a self-attention mechanism and a feedforward neural network. The self-attention mechanism allows each position in the encoder to attend to all positions in the previous layer of the encoder. The decoder, similar in structure to the encoder, generates the output sequence. It also contains a self-attention mechanism, but includes an additional attention layer that focusses on the encoder's output. This design enables the decoder to consider the entire input sequence when producing each element of the output. Transformers offer parallel processing of sequences, which significantly speeds up training and inference times. They have achieved state-of-the-art results in various NLP tasks, including translation, text summarisation, and question-answering. Their architecture has also inspired adaptations in other domains, such as computer vision. Building on the Transformer architecture, OpenAI introduced GPT (Generative Pretrained Transformer) and its subsequent iterations, which demonstrated remarkable language generation capabilities [25]. Similarly, Google's BERT (Bidirectional Encoder Representations from Transformers) model [24] brought advances in language understanding, especially in tasks like question answering and sentiment analysis. The current landscape of LLMs is characterised by their increasing size and sophistication, with models like GPT-3 and GPT-4 pushing the boundaries of what is possible in natural language processing and generation [28].

One of the most striking features of modern machine learning models, particularly in the domain of natural language processing, is their unprecedented size. Large Language Models (LLMs) have grown to colossal scales with a staggering number of parameters that defy conventional expectations. These parameters represent the learnt knowledge and patterns within the model. The scale of LLMs is often measured in billions or even trillions of parameters. For instance, models such as GPT-3 by OpenAI boast 175 billion parameters, dwarfing their predecessors by orders of magnitude. The growth of LLMs over the past few years has followed an exponential trajectory. From models with millions of parameters to those with billions and beyond, this rapid expansion has pushed the boundaries of what was previously thought possible. Each new iteration of LLMs pushes the envelope further, breaking records in terms of both performance and parameter count. Training and fine-tuning these gigantic models require an immense amount of computational resources. Training a model with trillions of parameters requires not only powerful GPUs or TPUs (Tensor Processing Units) but also extensive memory capacity and high-speed network connections. This computational demand has led to the development of specialised hardware and distributed training setups. Despite their enormous size, LLMs have found applications in a wide range of domains. They excel in natural language understanding and generation tasks, including translation, summarisation, question answering, and content generation. Their immense knowledge base allows them to generate coherent and contextually relevant text across diverse topics.

I would be guilty of neglecting important parts of the history of machine learning if I did not at least mention something about generative AI. Probably the most important contribution to generative AI was the introduction of Generative Adversarial Networks (GANs) by Goodfellow *et al.* [29], which marked a significant leap in the field [29]. GANs revolutionised the way machines could generate realistic images, leading to a surge in research and applications of generative models. The development of variational autoencoders (VAE) by Kingma *et al.* [30], published in 2013, explored a statistical method, offering a new approach to generative modelling [30]. These milestones collectively represent the rapid and ongoing evolution of generative AI, charting the path from simple pattern recognition to the creation of rich, complex and diverse outputs that can mimic human creativity.

Generative AI has become an important tool in synthetic data generation within scientific research. This approach enables scientists to create large, realistic datasets that can be used to train machine learning models, especially in fields where real data are scarce or difficult to obtain. By generating high-quality synthetic data that closely mirror real-world conditions, generative AI can facilitate more robust and extensive experimentation, significantly advancing research in areas like healthcare, environmental science, and materials engineering.

2.2 Machine Learning in Science

The previous section gave a brief history of machine learning from its inception. However, since this book is tailored to scientists, it is necessary to briefly highlight some of the major achievements of machine learning in various scientific disciplines.

In chemistry, ML has revolutionised several areas, notably drug discovery and material science. The traditional drug discovery process is often long and expensive. ML algorithms have significantly accelerated this process by predicting the properties and activities of molecules. For example, ML models can quickly screen vast chemical spaces for potential drug candidates, a process highlighted in the work of Gomez *et al.* [31]. Furthermore, ML has been instrumental in materials science for predicting the properties of new materials, thereby guiding experimental efforts in a more focused and efficient manner. Physics, especially high-energy physics and astrophysics, has benefited immensely from ML. In high-energy physics, ML methods have been crucial in analysing data from particle accelerators. The detection of the Higgs boson, as explained in [32], is a prime example in which deep learning significantly improved the separation of signals from background noise. In astrophysics, ML assists in interpreting vast amounts of data from telescopes and space missions, aiding in the discovery of new celestial objects and phenomena. In biology, perhaps the most notable contribution of ML is in the field of genomics and proteomics. Protein structure prediction, a long-standing challenge in biology, was revolutionised by the AlphaFold system developed by Google DeepMind [33]. The ability of this system to predict protein structures accurately and quickly has profound implications for understanding biological processes and designing new therapeutics. Furthermore, ML is increasingly used in genomics to understand genetic variations and their links to diseases. The integration of machine learning into scientific research not only has accelerated discoveries, but also has opened new avenues for exploration and innovation. As ML technology continues to evolve, its role in the advancement of scientific knowledge is expected to grow even further.

2.3 Types of Machine Learning

Now we need to focus on some terminology that is widely used in the machine learning community. The field of machine learning is vast, and there are many *types* of algorithms. It is important to shed some light on the various types to clarify what you will find in this book. A very broad and rough classification of ML approaches is given below.

- **Supervised Learning:** This method, fundamental to machine learning, involves training a model on a labelled dataset, where each example is paired with some expected output. The model learns to predict outcomes based on these data, using examples to infer patterns.
- **Unsupervised Learning:** Unlike supervised learning, unsupervised learning involves training models on data without predefined labels. It focusses on identifying hidden structures and patterns. Historically significant for its role in clustering and association problems, unsupervised learning is key in exploratory data analysis and dimensionality reduction techniques like Principal Component Analysis (PCA).
- **Semi-Supervised Learning:** This approach combines elements of supervised and unsupervised learning. It is particularly useful when labelled data are limited or costly to obtain, a common challenge in machine learning. By utilising a mix of a small amount of labelled data and a larger pool of unlabelled data, models in semi-supervised learning can achieve higher accuracy than unsupervised methods alone.
- **Reinforcement Learning:** This type of learning is distinguished by its focus on making sequences of decisions. The algorithm, often called an agent, learns to achieve a goal in an uncertain and potentially complex environment. Historically, reinforcement learning has roots in game theory and optimal control theory.

In this book, our focus will be on the mathematical foundations applicable to all methods, but our examples will primarily be drawn from supervised and unsupervised learning problems, which are predominantly utilised in scientific research projects. We will not delve into reinforcement learning or semi-supervised learning, as these are less frequently employed in the scientific domain.

References

1. Alan M Turing. Computing machinery and intelligence. *Mind*, 59(236):433–460, 1950.
2. Frank Rosenblatt. The perceptron: A probabilistic model for information storage and organization in the brain. *Psychological review*, 65(6):386, 1958.
3. Marvin Minsky and Seymour Papert. *Perceptrons: An Introduction to Computational Geometry*. MIT Press, Cambridge, MA, 1969.
4. J. Ross Quinlan. Induction of decision trees. *Machine learning*, 1(1):81–106, 1986.
5. David E Rumelhart, Geoffrey E Hinton, and Ronald J Williams. Learning representations by back-propagating errors. *Nature*, 323(6088):533–536, 1986.
6. Terrence J Sejnowski and Charles R Rosenberg. Parallel networks that learn to pronounce english text. *Complex systems*, 1(1):145–168, 1987.
7. Yann LeCun et al. Handwritten digit recognition with a back-propagation network. *Advances in neural information processing systems*, 2:396–404, 1990.
8. Yann LeCun, Léon Bottou, Yoshua Bengio, and Patrick Haffner. Gradient-based learning applied to document recognition. *Proceedings of the IEEE*, 86(11):2278–2324, 1998.
9. Corinna Cortes and Vladimir Vapnik. Support-vector networks. *Machine learning*, 20(3):273–297, 1995.
10. Vladimir N Vapnik and Alexey Ya Lerner. A class of algorithms for learning pattern recognition. *Automation and Remote Control*, 24(6), 1964.
11. Bernhard E Boser, Isabelle M Guyon, and Vladimir N Vapnik. A training algorithm for optimal margin classifiers. In *Proceedings of the fifth annual workshop on Computational learning theory*, pages 144–152. ACM, 1992.
12. Yann LeCun, Yoshua Bengio, and Geoffrey Hinton. Deep learning. *nature*, 521(7553):436–444, 2015.
13. David M Blei, Andrew Y Ng, and Michael I Jordan. Latent dirichlet allocation. *Journal of Machine Learning Research*, 3:993–1022, 2003.
14. Yoav Freund and Robert E Schapire. A short introduction to boosting. *Journal-Japanese Society For Artificial Intelligence*, 14:771–780, 1999.
15. Jerome H Friedman. Greedy function approximation: A gradient boosting machine. *Annals of statistics*, pages 1189–1232, 2001.
16. Richard S Sutton and Andrew G Barto. *Reinforcement learning: An introduction*. MIT press, 2018.
17. David Silver, Aja Huang, Chris J Maddison, Arthur Guez, Laurent Sifre, George van den Driessche, Julian Schrittwieser, Ioannis Antonoglou, Veda Panneershelvam, Marc Lanctot, et al. Mastering the game of go with deep neural networks and tree search. *nature*, 529(7587):484–489, 2016.
18. Richard S Sutton. Learning to predict by the methods of temporal differences. *Machine learning*, 3(1):9–44, 1988.
19. Gerald Tesauro. Temporal difference learning and td-gammon. *Communications of the ACM*, 38(3):58–68, 1995.
20. Richard S Sutton, David A McAllester, Satinder P Singh, and Yishay Mansour. Policy gradient methods for reinforcement learning with function approximation. *Advances in neural information processing systems*, 12:1057–1063, 2000.
21. Peter Auer, Nicolò Cesa-Bianchi, and Paul Fischer. Finite-time analysis of the multiarmed bandit problem. *Machine learning*, 47(2-3):235–256, 2002.
22. David Silver, Julian Schrittwieser, Karen Simonyan, Ioannis Antonoglou, Aja Huang, Arthur Guez, Thomas Hubert, Lucas Baker, Matthew Lai, Adrian Bolton, et al. Mastering the game of go without human knowledge. *Nature*, 550(7676):354–359, 2017.
23. Alex Krizhevsky, Ilya Sutskever, and Geoffrey E Hinton. Imagenet classification with deep convolutional neural networks. In *Advances in neural information processing systems*, pages 1097–1105, 2012.
24. Jacob Devlin, Ming-Wei Chang, Kenton Lee, and Kristina Toutanova. Bert: Pre-training of deep bidirectional transformers for language understanding. *arXiv preprint arXiv:1810.04805*, 2018.

25. Alec Radford, Karthik Narasimhan, Tim Salimans, and Ilya Sutskever. Improving language understanding by generative pre-training. 2018.
26. Ilya Sutskever, Oriol Vinyals, and Quoc V Le. Sequence to sequence learning with neural networks. *Advances in neural information processing systems*, 27, 2014.
27. Ashish Vaswani, Noam Shazeer, Niki Parmar, Jakob Uszkoreit, Llion Jones, Aidan N Gomez, Łukasz Kaiser, and Illia Polosukhin. Attention is all you need. *Advances in neural information processing systems*, 30, 2017.
28. Tom B Brown, Benjamin Mann, Nick Ryder, Melanie Subbiah, Jared Kaplan, Prafulla Dhariwal, Arvind Neelakantan, Pranav Shyam, Girish Sastry, Amanda Askell, et al. Language models are few-shot learners. *arXiv preprint arXiv:2005.14165*, 2020.
29. Ian J. Goodfellow et al. Generative adversarial nets. In *Advances in neural information processing systems*, pages 2672–2680, 2014.
30. Diederik P Kingma and Max Welling. Auto-encoding variational bayes. *arXiv preprint arXiv:1312.6114*, 2013.
31. Rafael Gómez-Bombarelli, Jennifer N Wei, David Duvenaud, José Miguel Hernández-Lobato, Benjamín Sánchez-Lengeling, Dennis Sheberla, Jorge Aguilera-Iparraguirre, Timothy D Hirzel, Ryan P Adams, and Alán Aspuru-Guzik. Automatic chemical design using a data-driven continuous representation of molecules. *ACS central science*, 4(2):268–276, 2018.
32. Pierre Baldi, Peter Sadowski, and Daniel Whiteson. Enhanced higgs boson to $\tau+$ $\tau-$ search with deep learning. *Physical review letters*, 114(11):111801, 2015.
33. Andrew W. Senior et al. Improved protein structure prediction using potentials from deep learning. *Nature*, 577:706–710, 2020.

Chapter 3
Calculus and Optimisation for Machine Learning

In the fall of 1972, President Nixon announced that the rate of increase of inflation was decreasing. This was the first time a president used the third derivative to advance his case for reelection.

Hugo Rossi

This chapter delves into the fundamental concepts of calculus and optimisation related to machine learning, offering both theoretical insights and practical use-cases. Starting with the motivation behind using calculus in machine learning, the chapter systematically introduces the concept of limit, which lays the foundation for understanding derivatives and their properties. The discussion on derivatives extends to their role in partial differentiation and gradients, both crucial for optimising machine learning algorithms. A significant portion of the chapter is dedicated to optimisation techniques specifically tailored for neural networks. The chapter begins with an overview of learning definitions and their implications for neural network training. This is followed by an examination of constrained versus unconstrained optimisation and an exploration of the complexities in identifying absolute and local minima of functions. The latter part of the chapter is focused on various optimisation algorithms, briefly discussing line search and trust region methods, then transitioning into specific approaches like steepest descent and gradient descent. The importance of selecting an appropriate learning rate is discussed, along with variations of gradient descent (GD) and strategies for choosing the right mini-batch size. The chapter concludes by exploring the connection between Stochastic Gradient Descent (SGD), fractals, and their implications in machine learning, providing a fascinating example of how complexity appear from very easy optimisation problems.

3.1 Motivation

Calculus is the branch of mathematics that concerns itself with the study of concepts that involves, among a great number of other things, infinitely small quantities (such as the rate of change of a function[1]) or infinitely large number of quantities (for

[1] If you do not know why the calculation of the rate of change of a function does involve infinitely small quantities hang on, it will become clear later in this chapter.

U. Michelucci, *Fundamental Mathematical Concepts for Machine Learning in Science*,
https://doi.org/10.1007/978-3-031-56431-4_3

example the sum of infinite terms in a series)[2]. It is one of the key components of deep learning (and in general machine learning), since it is at the core of the algorithms that are responsible for the training of models. The most important concepts in calculus that are relevant for deep learning are the following.

1. **Derivatives** (in one or many dimensions) are relevant for minimisation algorithms, the ones responsible for training models (as neural networks for example, but not only).
2. **Properties of Derivatives** are fundamental for understanding backpropagation[1] (the most important property is the *composition property*). Backpropagation is a fundamental algorithm used in training artificial neural networks, that is able adjust efficiently the weights during network training.
3. **Extrema of functions** (minima and maxima) and where to find them[3]. Note that there is an entire branch of applied mathematics called optimisation that deals with finding extrema. We will only scratch the surface of it in this book.

If you want to understand how optimisation algorithms (the ones responsible for minimisation of functions) work, you need to understand derivatives. We will start from there, then try to understand what properties extrema of functions have (and how to find them), and finally do a relatively deep and detailed study of the gradient descent algorithm, **the** most famous (but not the most used anymore) and instructive algorithm used to train deep neural networks and other algorithms.

3.2 Concept of Limit

Let us start with the concept of the limit of a function $f(x)$, which is necessary to understand derivatives. First, we need to define what a function is.

Definition 3.1 A function f from a set X to a set Y is a mapping from each element of X to exactly one element of Y. X is called the domain of the function and the set Y is called the codomain of the function.[3]

For simplicity we will consider only the case where $x \in \mathbb{R}$ and $f(x) \in \mathbb{R}$, or in other words $X = Y = \mathbb{R}$. An intuitive definition of a limit is as follows.

Definition 3.2 (intuitive) The limit of a function $f(x)$ in $x = x_0$ is the value L that the function approaches as the input x approaches some value x_0.

The limit L is typically indicated with the following notation.

$$L = \lim_{x \to x_0} f(x) \tag{3.1}$$

[2] That is probably the most inaccurate and generic description that is possible to give. I nonetheless hope that it can give some intuition of its applications.

[3] If you are a fan of Harry Potter you should get the pun.

In other (and still intuitive) words, Equation (3.1) means that we can make $f(x)$ as close to L as we want by choosing x close enough to x_0. For those of you who are more mathematically inclined, the formal definition of a limit is as follows.

Definition 3.3 (★ Formal Definition of Limit) L is called the limit of the function $f(x)$ in x_0 if given an arbitrary real number $\epsilon > 0$, there is a $\delta \in \mathbb{R}$ such that for any x that satisfies $|x - x_0| < \delta$, it is true that $|f(x) - L| < \epsilon$.

You may think that it is easy to calculate the limit of a function. Why do not simply calculate $f(x_0)$? In some cases it is really that simple. For example,

$$\lim_{x \to x_0} x^2 = x_0^2 \tag{3.2}$$

But what is the value of the following limit[4]?

$$\lim_{x \to 0} \frac{\sin x}{x} \tag{3.3}$$

You cannot simply calculate the value of $\sin x / x$ in 0, since the function is not defined. To calculate such limits, there are plenty of methods and tricks, but those go beyond the scope of this book. In machine learning applications, you will never need to calculate a limit. But the concept is important to understand derivatives.

Info ★ Geometric proof that $\lim_{x \to 0} \frac{\sin x}{x} = 1$

Since I do not like to give you any statement without proving it if possible, here is a proof of

$$\lim_{x \to 0} \frac{\sin x}{x} = 1. \tag{3.4}$$

based on geometry. Note that to understand it you need to have a basic understanding of trigonometry.

Proof Consider Figure (3.1).

[4] The value of this limit is 1, but it is difficult to calculate, unless you know derivatives and advanced methods as the l'Hopital rulefor limits.

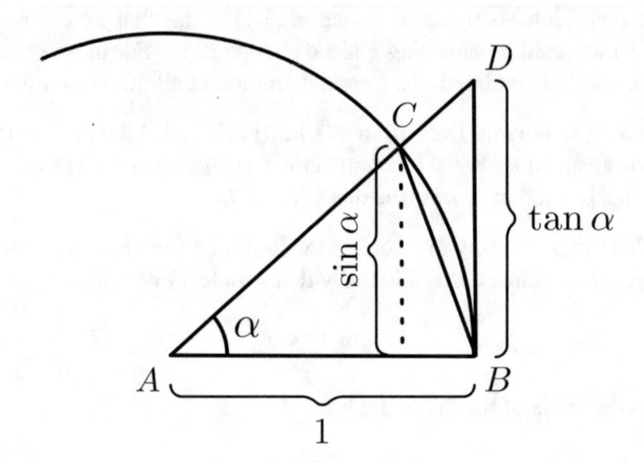

Fig. 3.1 Geometrical construction to justify the proof of $\lim_{x \to 0} \sin x / x = 1$.

The area of the triangle ABC (A(ABC)) is

$$A(\text{ABC}) = \frac{1}{2} \sin \alpha. \tag{3.5}$$

The circle wedge limited by the letters ABC is $A_{\text{wedge}}(\text{ABC}) = \alpha/2$. Recall that this comes from the fact that the angle α is expressed in radians. The area of the triangle A(ABD) is

$$A(\text{ABD}) = \frac{1}{2} \tan \alpha. \tag{3.6}$$

Now, from Figure (3.1) you can see that

$$A(\text{ABD}) < A_{\text{wedge}}(\text{ABD}) < A(\text{ABC}) \tag{3.7}$$

that transaltes into

$$\frac{1}{2} \tan \alpha < \frac{\alpha}{2} < \frac{1}{2} \sin \alpha \tag{3.8}$$

dividing by $1/2 \sin \alpha$ and taking reciprocals we get

$$\cos x < \frac{\sin x}{x} < 1 \tag{3.9}$$

Now since

$$\lim_{x \to 0} \cos x = 1 \tag{3.10}$$

from the disequalities in Equation (3.9) and by taking the limit for $x \to 0$ it follows that

$$\lim_{x \to 0} \frac{\sin x}{x} = 1. \tag{3.11}$$

3.3 Derivative and its Properties

Derivatives are essential in machine learning and deep learning because they are the key to optimisation algorithms, including the backpropagation algorithm in deep learning.

The derivative can be intuitively understood in one dimension. Let us consider a generic function $y = f(x)$ with $x \in \mathbb{R}$ and $y \in \mathbb{R}$. The derivative of $f(x)$ at a point x_0 is indicated by the symbols

$$f'(x_0) \text{ or } \frac{df}{dx}(x_0) \tag{3.12}$$

and it can be interpreted geometrically as the slope of the tangent of the function at $x = x_0$. In Figure 3.2 you can see an example. The function $f(x) = x^2$ is plotted in black and the tangent to $f(x)$ for $x = x_0 = 3$ is plotted in red. The slope of the tangent is the derivative of $f(x)$ at $x = x_0 = 3$. The derivative is defined in terms of

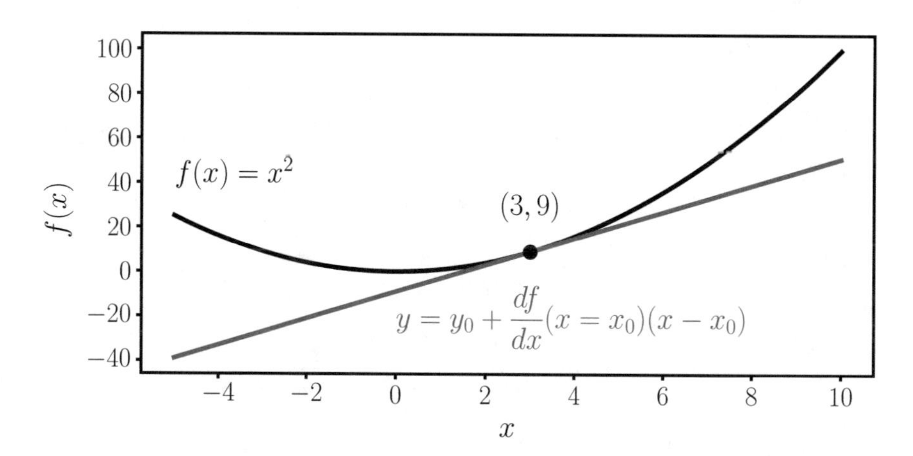

Fig. 3.2 When dealing with a function $f(x)$ of a one-dimensional variable x, the derivative at a given point x_0 is the slope of the tangent to the curve at the point $(x_0, f(x_0))$. The red curve in the figure is the tangent at the point $(3, 9)$ of the function $f(x) = x^2$. The slope of the red line is the derivative of the function (x) evaluated at 3. In this case $f'(x) = 2x$ and $f'(3) = 6$.

a limit. Formally, one can give the following definition.

Definition 3.4 A function of $f(x) : \mathbb{R} \to \mathbb{R}$ is differentiable at a point x_0 of its domain[5], if its domain contains an open interval[6] I containing x_0, and the limit

[5] The domain of a function is the set of all inputs accepted by the function.

[6] Discussing what an open interval is goes beyond the scope of this book, so you can ignore it. I included it because I wanted to give a precise definition.

$$L = \lim_{\Delta x \to 0} \frac{f(x + \Delta x) - f(x)}{\Delta x} \tag{3.13}$$

exist. In this case L is called the derivative of $f(x)$ at $x = x_0$ and $L = f'(x_0)$.

Info ★ Existence of the Limit

A limit of a function $g(x)$ for $x \longrightarrow x_0$

$$q = \lim_{x \to x_0} g(x) \tag{3.14}$$

is said to exist if for every positive real number ϵ, there exists a positive real number δ such that $0 < |x - x_0| < \delta$ implies $|g(x) - q| < \epsilon$. In other words, it means that if x is close to x_0, then $g(x)$ will be close to q.
For example, let us consider the function x^2 and let us prove the following lemma.

Lemma 3.1 *The limit*

$$L = \lim_{x \to x_0} x^2 \tag{3.15}$$

always exist for all finite x_0 *and is* $L = x_0^2$.

Proof Let us consider an $\epsilon \in \mathbb{R}$. Let us find out what δ we need to choose such that from

$$|x - x_0| < \delta \tag{3.16}$$

it follows

$$|x^2 - x_0^2| < \epsilon. \tag{3.17}$$

Let us first note that

$$|x - x_0| < \delta \Rightarrow x < \delta + x_0 \text{ or } x < x_0 - \delta \tag{3.18}$$

By using Equation 3.18 we can write

$$x < \delta + x_0 \Rightarrow x + x_0 < \delta + 2x_0 \tag{3.19}$$

now we are almost there. In fact, now we can write

$$|x^2 - x_0^2| = |x - x_0||x + x_0| < \delta|x + x_0| < \delta(2x_0 + \delta) \tag{3.20}$$

so to satisfy $|x^2 - x_0^2| < \epsilon$ we simply need to choose a δ given by the equation $\delta(2x_0 + \delta) = \epsilon$, that have the solution $\delta = \sqrt{x_0^2 + \epsilon} - x_0$. □

Lemma 3.1 can be better understood by looking at Figure 3.3.

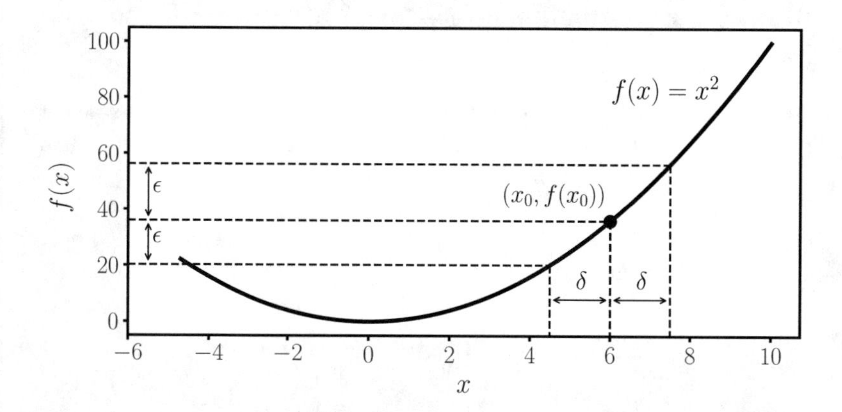

Fig. 3.3 In the figure you can see how, for the function x^2, for any ϵ chosen (the vertical intervals), there is always a δ that corresponds to an interval in x. In the figure we have chosen $x_0 = 6$. This is a graphical explanation of the concept of the existence of a limit. In fact, one can show that $\lim_{x \to x_0} x^2$ is always existing for any finite x_0 as we proved in Lemma 3.1.

In this book we will not concern ourselves with the problem of existence of derivatives, but you should be aware that the derivative is not always existing in the domain of a function. Consider, for example, the function $f(x) = \max\{0, x\}$. This is called the Rectified Linear Unit (ReLU) and is widely used in deep learning. But this function does not have a derivative at $x = 0$. In fact, if you plot it, you will see that at x_0 the function has no clear and definite slope. In Figure 3.4 you can see a plot of the ReLU function.

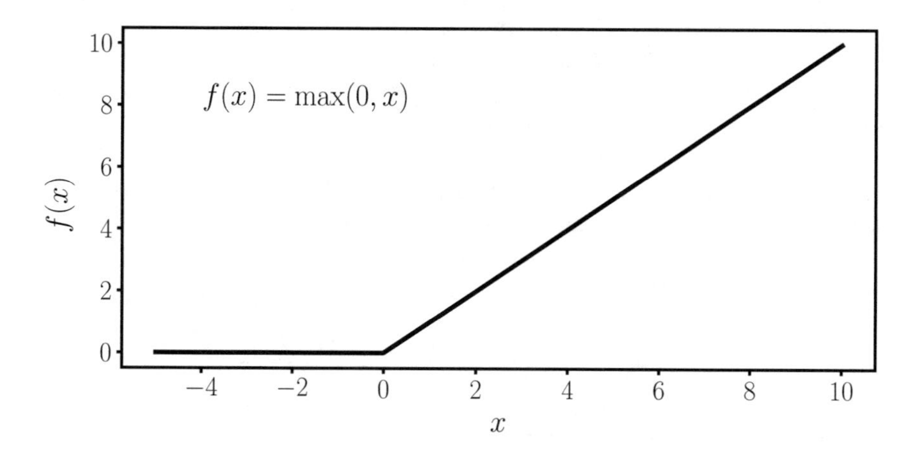

Fig. 3.4 The function ReLU ($\max\{0, x\}$, plotted in this figure, has no derivative at $x = 0$ as can be seen intuitively from the plot.

Warning ReLU function missing derivative at $x = 0$ and deep learning

We have just seen that the derivative of the ReLU function is not defined at $x = 0$. So how comes that this function is so used in deep learning when training neural networks? Surely this must be a problem!
In the numerical world, things are slightly different than in theory. When implementing the derivative of the ReLU function, it is enough to simply assign a specific value at $x = 0$. It is enough to program it in the code as

$$\frac{d\text{ReLU}}{dx} = \begin{cases} 0 & \text{for } x < 0 \\ 0(\text{or } 1) & \text{for } x = 0 \\ 1 & \text{for } x > 0 \end{cases} \tag{3.21}$$

Note that this is, strictly speaking, equal to the derivative of the ReLU function only for $x \neq 0$, but this makes the numerical routines work. Additionally, due to rounding errors in Python (the same applies to other programming languages), the output of a calculation between floating point variables will, for all practical purposes, never be **equal** to zero, making this discussion moot.
The Google TensorFlow library goes a step beyond that. It simply uses the derivative of the ReLU function only for $x > 0$, since if the derivative is zero, it is useless anyway. But it is important to be aware of such problems, for example, when you want to develop your own activation or loss functions.
If you create in Python a numpy array with one element as the `float16` datatype, for example with the line `a = np.array([0.1], dtype = np.float16)`, the output of printing it will be `0.0999755859375`, showing how rounding errors are always present due to the way numbers are stored in the memory of computers. Some exceptions apply, but we will not spend any more time on this. Just be aware that derivatives are not always defined.

In general, you do not have to calculate derivatives with limits. By using fundamental derivatives that someone calculated for you and by using their properties (see next section), you can easily calculate the derivative of almost all functions. Just as an example, let us see how to calculate the derivative of x^2. By using definition 3.13 with $f(x) = x^2$ we have

$$\begin{aligned} \frac{df(x)}{dx} &= \lim_{\Delta x \to 0} \frac{f(x+\Delta x) - f(x)}{\Delta x} = \lim_{\Delta x \to 0} \frac{(x+\Delta x)^2 - x^2}{\Delta x} = \\ &= \lim_{\Delta x \to 0} \frac{\Delta x^2 + 2x\Delta x}{\Delta x} = \\ &= \lim_{\Delta x \to 0} (\Delta x + 2x) = 2x \end{aligned} \tag{3.22}$$

So the derivative of x^2 is $2x$. It is possible (and not difficult) to generalise this result. The derivative of x^r is, in fact, rx^{r-1} for r a real number. The proof is easier for r integer and can be done analogously with what I have done here. You can find easily tables of derivatives of many functions, ranging from exponentials, logarithms, and many more.

3.3.1 Derivatives' Properties

Often, you'll find yourself needing to compute derivatives of complex functions that are composed of simpler ones. For example you may want to calculate the derivative of

$$e^x + x^2 \tag{3.23}$$

or

$$\cos(\tan(e^x)). \tag{3.24}$$

It is very useful to know how derivatives behave when dealing with composition of functions. Here is a list of properties that you will find useful.

$$\begin{aligned}
(f(x) + g(x))' &= f'(x) + g'(x) \\
(f(x) - g(x))' &= f'(x) - g'(x) \\
(f(x)g(x))' &= f'(x)g(x) + f(x)g'(x) \\
\left(\frac{f(x)}{g(x)}\right)' &= \frac{f'(x)g(x) - f(x)g'(x)}{g(x)^2}
\end{aligned} \tag{3.25}$$

There is one additional property that is the basis for how backpropagation works. Backpropagation is at the core of how it is possible to calculate derivatives of complicate neural network functions. This property tells you how to calculate the derivative of the composition of functions.

$$(g(f(x)))' = g'(f(x))f'(x) \tag{3.26}$$

The composition of functions is often indicated with

$$(g \circ f)\,(x) = g(f(x)) \tag{3.27}$$

Info ★ Proof of the Formula $(g(f(x)))' = g'(f(x))f'(x)$

Proof The formula can be proven by using the definition of a derivative.

$$
\begin{aligned}
(g \circ f)'(x) &= \lim_{h \to 0} \frac{(g \circ f)(x+h) - (g \circ f)(x)}{h} \\
&= \lim_{h \to 0} \frac{g(f(x+h)) - g(f(x))}{h} \\
&= \lim_{h \to 0} \frac{g(f(x+h)) - g(f(x))}{h} \times \frac{f(x+h) - f(x)}{f(x+h) - f(x)} \\
&= \lim_{h \to 0} \frac{g(f(x+h)) - g(f(x))}{f(x+h) - f(x)} \times \frac{f(x+h) - f(x)}{h} \\
&= \lim_{h \to 0} \frac{g(f(x+h)) - g(f(x))}{f(x+h) - f(x)} \times \lim_{h \to 0} \frac{f(x+h) - f(x)}{h} \\
&= \lim_{h \to 0} \frac{g(f(x+h)) - g(f(x))}{f(x+h) - f(x)} \times f'(x)
\end{aligned}
\tag{3.28}
$$

by doing the change of variable $k = f(x+h) - f(x)$

$$
\begin{aligned}
k &= f(x+h) - f(x) \\
f(x+h) &= f(x) + k
\end{aligned}
\tag{3.29}
$$

one can see that

$$
\lim_{h \to 0} k = \lim_{h \to 0} f(x+h) - f(x) = f(x+0) - f(x) = 0 \tag{3.30}
$$

by using k we can write

$$
\begin{aligned}
\lim_{h \to 0} \frac{g(f(x+h)) - g(f(x))}{f(x+h) - f(x)} &= \lim_{k \to 0} \frac{g(f(x)+k) - g(f(x))}{k} \\
&= g'(f(x))
\end{aligned}
\tag{3.31}
$$

and thus finally

$$
\begin{aligned}
(g \circ f)'(x) &= \lim_{h \to 0} \frac{g(f(x+h)) - g(f(x))}{f(x+h) - f(x)} \times f'(x) \\
&= g'(f(x)) \cdot f'(x)
\end{aligned}
\tag{3.32}
$$

But why is this property so important in deep learning? The reason lies in the fact that in practically all kind of neural network architectures, the output is obtained by calculating the composition of a large (in some cases up to as much as 50 or more) number of functions. A network, in its most simple form, is a sequential stack of layers. Each layer takes the output of the preceding layer as its input. This means that the final layer is a function of the one before it, the second-to-last layer is a function of the third-to-last, etc. Consequently, the output of your neural network is the result of multiple function compositions, resembling something like

$$f_1(f_2(f_3(...f_N(x))))...) \tag{3.33}$$

with N some large number and $f_i(x)$ some complicated and non-linear function[7]. Even if you still do not know, to train a neural network, you have to calculate the derivative of such functions. Therefore, property (3.26) is of fundamental importance. Note that when implementing neural networks, you do not have to calculate derivatives yourself. Libraries like TensorFlow or pyTorch will do it for you. In the background, they still use the *composition property* in a smart form, called, as I mentioned, autodifferentation and backpropagation.

3.4 Partial Derivative

The next concept you will need, especially in deep learning, is that of a partial derivative. This becomes relevant when you have a function of a variable $\mathbf{x} \in \mathbb{R}^n$, with $n > 1$. Take, for example, a function $f(x_1, x_2, ..., x_n)$. The partial derivative of f with respect to x_i in $(x_1, x_2, ..., c, ..., x_n)$ (with $x_i = c$) is indicated with

$$\frac{\partial f(x_1, ..., x_n)}{\partial x_i}(x_i = c) \tag{3.34}$$

and is defined by (note that the value c is at the ith position)

$$\frac{\partial f(x_1, ..., x_n)}{\partial x_i} = \lim_{h \to 0} \frac{f(x_1, ..., c + h, ..., x_n) - f(x_1, ..., c, ..., x_n)}{h} \tag{3.35}$$

if the limit exists of course. This is nothing else than the slope of the function f along the ith direction. It is called *partial* since the derivative is calculated by varying only x_i and keeping all other inputs constant. In Figure (3.5) you can see the 3-dimensional plot of a function $g(x, y)$. The line indicated with L_x is the tangent to the surface at the point P along the x axis (its slope is the partial derivative of $g(x, y)$ with respect to x), and the line indicated with L_y is the tangent to the surface at point P along the y axis (its slope is the partial derivative of $g(x, y)$ with respect to y).

3.5 Gradient

Let us consider a scalar function of multiple variables $f(x_1, x_2, ..., x_n)$. The gradient of f is indicated with ∇f and is defined by

[7] In a feed-forward neural network for example, the f_i may be interpreted as the layer functions of the network.

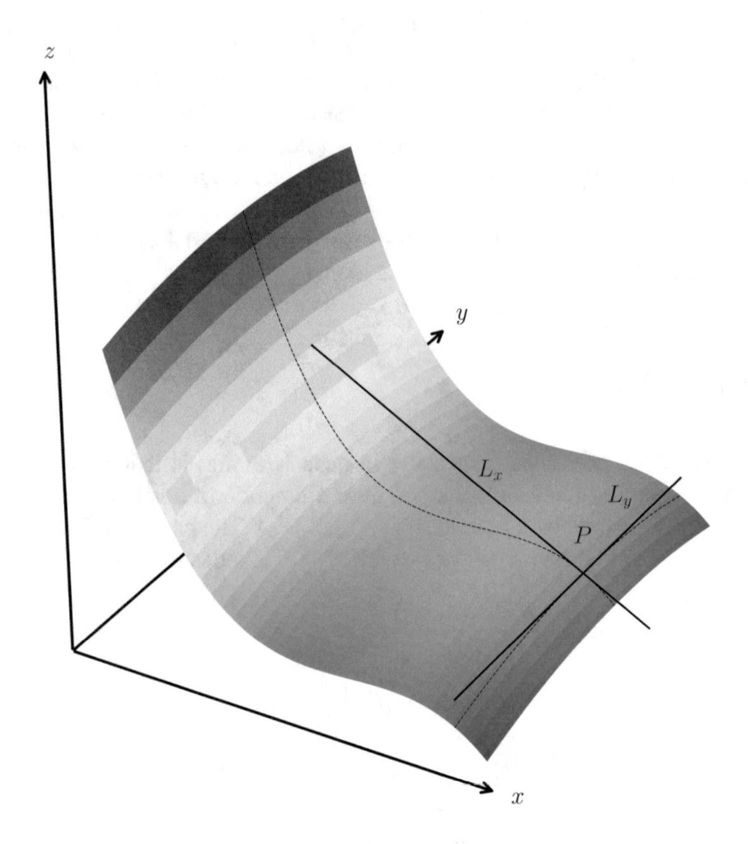

Fig. 3.5 In the figure you can see the 3-dimensional plot of a function $g(x, y)$. The line indicated with L_x is the tangent to the surface at point P along the x-axis and its slope is $\partial g/\partial x(P)$, and the line indicated with L_y is the tangent to the surface at point P along the y axis, and its slope is $\partial g/\partial y(P)$.

$$\nabla f(x_1, ..., x_n) = \begin{pmatrix} \frac{\partial f}{\partial x_1} \\ \cdots \\ \frac{\partial f}{\partial x_n} \end{pmatrix} \tag{3.36}$$

that is, a vector with partial derivatives along the directions x_1, x_2, and so on. We will limit ourselves here to giving only the definition, and you will see how it is used in this book more extensively. If you do not know what a vector is, for now it is sufficient to know that this is simply an array of many elements. In Chapter 4 we will discuss this concept in detail.

3.6 Extrema of a Function

Derivatives are key in optimisation to understand how functions behave, which is essential when seeking optimal solutions. The objective in optimisation is typically to find the highest or lowest value that a function can achieve. The derivatives signal where these extremes, i.e., maxima or minima, might be located. The first derivative, indicating the function's rate of change, helps identify critical points where the function's slope is zero, suggesting potential maxima and minima. The second derivative further clarifies the nature of these critical points by differentiating between a maximum, a minimum, or a saddle point (in multiple dimensions) or an inflection point (in one dimension).

Now, let us dive into what exactly are *extreme* points of a function. **Extreme points** refer to places where a function reaches a local or global maximum, minimum or an inflection point. Visually, these are the points on a function's graph that correspond to peaks, valleys, and saddles, signifying a change in the function's direction from increasing to decreasing, or the reverse. It is important to properly define the types of extreme point we encounter. Considering a function $f : \mathbb{R} \to \mathbb{R}$, I find that while the definitions for multi-dimensional functions are similar, they are often more easily understood in the one-dimensional case. This is why we will consider only the one-dimensional case here in this section.

- **Global Maximum or Minimum**: a function has a global maximum or minimum in x_0 if $\forall x \in \mathbb{R}\ f(x) \leq f(x_0)$ or $f(x) \geq f(x_0)$ respectively. Imagine a landscape with various peaks and valleys. The global maximum would be the highest peak in this entire landscape, higher than all other peaks. The global minimum would be the lowest valley (see the plots on the left in Figure 3.6).
- **Local Maximum or Minimum**: a function has a local maximum or minimum in x_0 if $\exists \epsilon$ such that $\forall x \in [x_0-\epsilon, x_0+\epsilon]\ f(x) \leq f(x_0)$ or $f(x) \geq f(x_0)$ respectively. A local minimum is a point where the function has a value that is lower than at nearby points, but not necessarily the lowest overall in the entire domain. Imagine that you are walking through a hilly landscape. A local minimum is like a small valley surrounded by higher hills. From your position in that valley, all immediate steps lead upward, so it seems like the lowest point in that immediate area. However, beyond those surrounding hills, there might be deeper valleys. A local maximum occurs when you are on top of a hill, but you see much higher peaks around you. The local minimum, for example, is a point where the function is at its minimum locally, but not necessarily the absolute lowest point or the "global minimum" of the entire landscape. In optimisation, especially with complex functions, finding the global minimum can be challenging because algorithms might get stuck in these local minima, thinking they have found the lowest point when they have only found the lowest point in a small region. A local maximum would be a hill flanked by the highest peaks (see the plots in the central column in Figure 3.6).
- **Inflection Point**: a function has an inflection point at x_0 if the second derivative of the function $f''(x)$ changes sign as x passes through x_0. In other words, x_0

is an inflection point if it has a sign for $x < x_0$ and the opposite for $x > x_0$. Intuitively, imagine that you are walking through a hilly landscape. You have an inflection point when, for example, you walk upwards, you find a flat spot where you can rest, and then if you continue in the same direction you continue to walk upwards. In graphs, inflection points are where the shape of the curve changes. They are essential for understanding the behaviour of a function, as they indicate transitions in the growth and decay patterns of the function (see plots on the right in Figure 3.6).

As we have discussed, all these points are collectively called **extreme points**.

Info ★ Minima, Maxima and Inflection Point Definition

To be precise the three kind of points, minima, maxima and inflection, are related to the derivatives in the following way.

- A point x_0 is a minimum of a twice differentiable function if $f'(x_0) = 0$ and $f''(x_0) > 0$.
- A point x_0 is a maximum of a twice differentiable function if $f'(x_0) = 0$ and $f''(x_0) < 0$.
- A point x_0 is an inflection point of a twice differentiable function if $f'(x_0) = 0$ and $f''(x_0) = 0$.

Now we mention only a *twice differentiable* function, since otherwise we cannot talk about a second derivative if the latter does not exist.

By calculating the first derivative of a function, we can determine where its slope is zero, which typically indicates potential maxima or minima. These points, where the derivative equals zero, are also called critical points. The second derivative can further be used to discern the nature of these critical points: a positive second derivative suggests a local minimum, while a negative second derivative indicates a local maximum. In essence, derivatives serve as powerful tools in pinpointing where a function reaches its highest or lowest values, essential for optimisation and analysis in various mathematical and practical contexts. In machine learning, as we will discuss at length in the next sections, our objective is to minimise the discrepancy between a model's output and the expected values. Ideally, we aim to locate the global minimum of this error function (which we have not yet defined or discussed), as it represents the optimal solution. However, during model training, the algorithms tasked with reducing the error function, known as numerical optimisers, often encounter local minima, particularly with intricate error functions such as those in neural networks. Consequently, optimisation remains a vibrant area of research. Developing optimisers that are efficient, fast, and adept at avoiding local minima is crucial to effectively training complex machine learning models.

Let us consider one simple example and find the extreme point of $f(x) = x^2$ (the plots on the left in Figure 3.6). We start by calculating the location of the extremes, by finding the solutions to the equation $f'(x) = 0$.

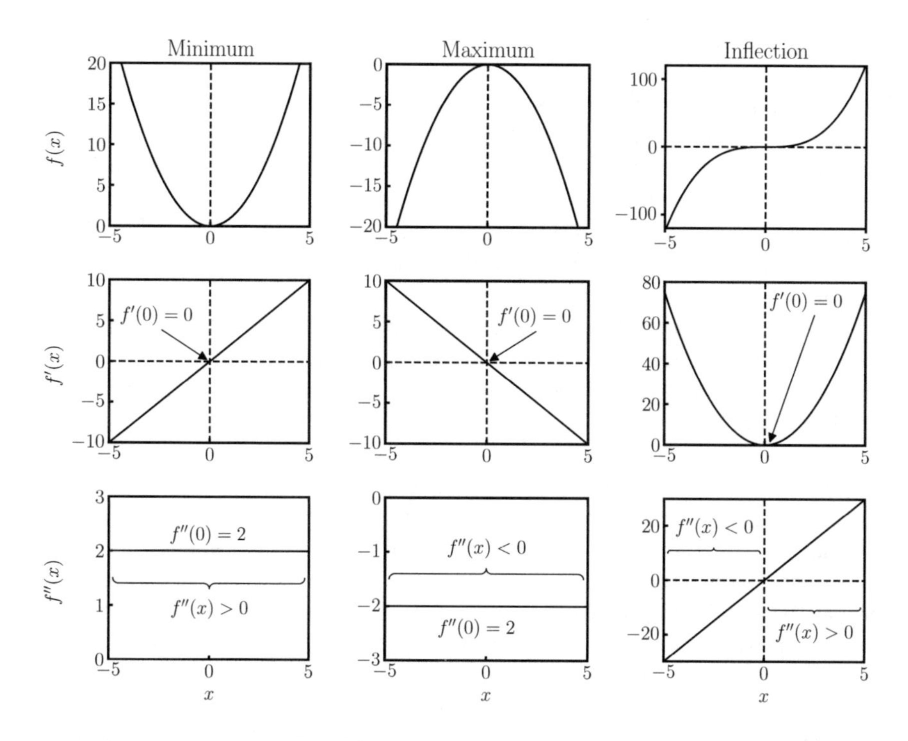

Fig. 3.6 In the left panel first row the function $f(x) = x^2$ is plotted. The point $x = 0$ is a **global minimum**. As can be seen in the second row $f'(0) = 0$ and in the third row $f''(x) > 0\ \forall x$, so that $x = 0$ is a global maximum. In the first row of the central panel, the function $f(x) = -x^2$ is plotted. The point $x = 0$ is a **global maximum**. As can be seen in the second row $f'(0) = 0$ and in the third row $f''(x) < 0\ \forall x$, so that $x = 0$ is a global maximum. In the right panel, first row, the function $f(x) = x^3$ is plotted. The point $x = 0$ is an **inflection point**. As can be seen in the second row $f'(0) = 0$ and in the third row $f''(x) < 0$ for $x < 0$ and $f''(x) > 0$ for $x > 0$, so that $x = 0$ is an inflection point.

$$f'(x) = 0 \Rightarrow 2x = 0 \Rightarrow x = 0 \tag{3.37}$$

Since the equation $f'(x) = 0$ has only a solution the function has one single extreme point x_0 that is $x_0 = 0$. Since $f''(0) > 0$, we can determine that x_0 is a minimum.

Now we need to discuss an important question before moving on. Let us indicate this phantomatic error function that measures the discrepancies between predictions and expectations for a model with $L(\theta)$, where we have indicated the model parameters that we want to optimise with θ (θ generally indicates a large number of parameters, not a single number). The entire game in machine learning consists of trying to find the values of the parameters θ_0 such that $L(\theta_0)$ has a global minimum. In other words,

$$\theta_0 = \arg\min_{\theta} L(\theta) \tag{3.38}$$

If you have followed the discussion so far, you may ask why don't we use the derivatives to find the best parameters? This would give us all the extreme points that we could check with the second derivatives. In theory this is correct but is not a practicable way. The three major problems that stop us from doing this are: the sheer number of parameters involved (do you remember for example that in GPT-3 we have trillions of parameters?), the complexity of the functions, and the fact that in many cases we don't even know how to write the functions in analytical form. Take, for example, neural networks. If you consider modern architectures, it is impossible to write an analytical form of the entire network, let alone calculating the derivatives analytically. The only possibility is to let a computer do it. In the case of neural networks, this is achieved with the help of two algorithms: backpropagation (which we have mentioned earlier while discussing the history of machine learning) and autodifferentiation.

For the sake of clarity, it is useful to briefly discuss those two concepts. **Backpropagation** is a fundamental algorithm used in the training of neural networks, particularly in the context of supervised learning. Mathematically, it involves the calculation of the gradient of a loss function with respect to the weights of the network. This algorithm uses the chain rule from calculus to propagate errors backward through the network, from the output layer to the input layer. The process consists of two main phases: a forward pass, where the input is passed through the network to obtain the output and compute the loss; and a backward pass, where the gradient of the loss is propagated back to update the weights, typically using a gradient descent algorithm (for a discussion see the next sections). The objective is to minimise the loss function, thereby improving the model's predictions. **Automatic Differentiation (AD)**, also known as algorithmic differentiation or computational differentiation, is a set of techniques used to numerically evaluate the derivative of a function specified by a computer program. AD takes advantage of the fact that every computer program, no matter how complex, executes a sequence of elementary arithmetic operations and functions (like addition, multiplication, sine, cosine, etc.). AD applies the chain rule of calculus repeatedly to these operations to compute derivatives of arbitrary order efficiently and accurately. Unlike symbolic differentiation, it does not suffer from expression swell (the problem that occurs when the formulas for a derivative grow and grow out of control until the point when they are not manageable anymore) and is more efficient and precise than numerical differentiation methods like finite differences. AD is particularly useful in machine learning to optimise machine learning models.

So even if we cannot use the simple formula $f'(x) = 0$ to find the minima of our error (or loss) functions, derivatives still play a key role in being able to train complex machine learning models. There are people who are trying to find alternatives, but there are few. Check, for example, the work by Kirsch and Schmidhuber [2]. The vast (and I should say everyone) majority of the machine learning community uses backpropagation and autodifferentiation and thus is dependent on derivatives to train their models. Both Python libraries TensorFlow and pyTorch use backpropagation.

Now that the role and importance of derivatives are clear, it should also be evident why calculus is such an important part of mathematics to master if you want to understand how machine learning works. In the next sections, we will discuss the most known optimiser: the gradient descent, after having discussed a bit more in detail what **learning** means in the context of machine learning and in particular of neural networks.

3.7 Optimisation for Machine Learning

Machine learning algorithms operate by minimising a specific function that evaluates the accuracy of the model's output compared to the expected result. This process involves continuously adjusting the model's parameters to minimise the difference or error between the predicted and actual outcomes. By doing so, the algorithm improves the model's ability to make accurate predictions or decisions, enhancing its overall performance in tasks such as classification, regression, or pattern recognition. The success of these algorithms largely depends on how effectively they can reduce the error and refine the model's predictions over the training process.

To grasp this method, let us consider the simplest example: linear regression. This problem is set up as follows: Given a data set that contains M pairs (x_i, y_i) where $i = 1, \ldots, M$, our goal is to determine the optimal values for the parameters a and b in equation $\hat{y} = ax + b$ so that the predictions ($\hat{y}_i = ax_i + b$) of the model are as close to the expected output y_i as possible. What docs *as close as* mean? We need to translate this into some mathematical form to be able to work with it. The classical approach is to define a mathematical function that measures how **far** the predictions are from the expected values (the **error**) and then minimise this function. Linear regression can be solved by minimising the mean square error (MSE) given by

$$\text{MSE} = \frac{1}{M}\sum_{i=1}^{M}(y_i - \hat{y}_i)^2 \tag{3.39}$$

where $\hat{y}_i = ax_i + b$. The MSE quantifies the model error. A higher MSE value indicates worse model performance. Optimisation focuses on strategies and theories that guide us in efficiently minimising MSE to find the best values for a and b.

Info Optimisation in Machine Learning

Optimisation plays a pivotal role in machine learning, serving as the foundation for effectively training models. It involves fine-tuning the parameters of a model to minimise errors and enhance accuracy, ensuring that the model's predictions are as close as possible to the actual outcomes. This process is crucial because it directly impacts a model's ability to learn from data and perform tasks like classification, regression, or pattern recognition with higher precision. Without effective optimisation techniques, machine learning models may struggle to converge to the best solution, leading to poor performance and limited practical applicability. Therefore, optimisation is not just a technical step but a fundamental aspect that drives the success and efficiency of machine learning algorithms in various applications.

This is a vast area of research that goes beyond the scope of this book. For the interested reader, here are two book suggestions: the classical convex optimisation text by Boyd and Vandenberghe [3] and the one on numerical optimisation by Nocedal and Wright [4]. Naturally there are many books on the subject, but those two are a valid starting point to understand the mathematical theory and applications of optimisation.

In the next section, we will discuss at length what optimisation means with some terminology that is relevant in the context of neural networks. Studying it in this context is particularly useful for grasping its general role in machine learning. Discussing optimisation in the case of neural networks will also allow us to define and understand the concept of mini-batch training or stochastic gradient descent that are otherwise not so relevant in the case of more classical machine learning algorithms such as decision trees, for example. Neural networks often involve a large number of parameters and layers, making the optimisation process more intricate and crucial to their performance. Mastering optimisation techniques in this challenging environment provides deep insights into how algorithms adjust to minimise errors and improve learning, insights that are applicable across various machine learning models. Furthermore, since neural networks are at the forefront of many advanced machine learning applications, understanding their optimisation paves the way for a comprehensive understanding of machine learning optimisation as a whole, including its challenges, strategies, and impacts on model effectiveness.

Info ★ Extreme Points in Many Dimensions

In our discussion we have only used examples in one dimension. But loss functions always have multiple parameters. Even a plain linear regression has two parameters. Consider a function with n parameters $g(x_1, \cdots, x_n)$. To find the extreme points of g we need to solve a system of n equations

$$\frac{\partial g}{\partial x_i} = 0 \text{ with } i = 1, ..., n \tag{3.40}$$

You may realize that, remembering from the definition of a gradient, that the previous set of equations can be written in vector form as

$$\nabla g = \mathbf{0} \tag{3.41}$$

where with $\mathbf{0}$ we have indicated a vector with all n components equal to zero. The role that the derivative has in one dimension in finding extreme points of a function, is taken by the gradient in multiple dimensions. This is, for example, the reasons why the most known optimisation algorithms is called *gradient* descent (as we will discuss later).

3.8 Introduction to Optimisation for Neural Networks

Although this book is not specifically about neural networks, it is nonetheless useful to discuss optimisation in the context of neural networks, since it is where it plays the major role. Concepts like mini-batch gradient descent (that we will discuss later in this chapter) are relevant only when dealing with neural networks for example. For this introductory section, we will consider only what is called supervised learning[8]. Suppose that we have a dataset of M tuples (x_i, y_i) with $i = 1, \ldots, M$. The x_i, called input observations or simply inputs, can be anything from images to multidimensional arrays, to one-dimensional arrays or even simple numbers. The outputs y_i (also called target variables or sometime labels) can be multidimensional arrays, numbers (for example, the probability of the input observation x_i of being of a specific class) or even images. In the most basic formulation, a neural network is a mathematical function (sometime called *network function*) that takes some kind of input (typically multidimensional) x_i (the subscript i indicates that we have a dataset of input observations at our disposal, and now we are considering only the ith one) and with it calculate some output $\hat{y}_i$. This function depends on a certain number N of parameters, that we will indicate with θ (that is a multidimensional array of values). We can write this mathematically as

$$\hat{y}_i \equiv f(\theta, x_i) \tag{3.42}$$

Where we have indicated the parameters in vector form $\theta = (\theta_1, \ldots, \theta_N) \in \mathbb{R}^N$. In Figure 3.7 you can see an intuitive diagram of the idea. The blob in the middle represents the network function f that maps the input x_i to the output. Naturally, the output will depend on the parameters θ. The idea behind learning is to change the parameters until $\hat{y}_i$ is as close to y_i as possible. There are two very important undefined concepts in the last sentence: firstly what "close" means, and secondly how can we update the parameters in an intelligent way to make $\hat{y}_i$ and y_i "close". We will answer those exact two questions in depth in this chapter.

To summarise, a neural network is nothing else than a mathematical function that depends on a set of parameters that are tuned, hopefully in some smart way, to make the network output as close as possible to some expected output. The concept of "close" is not defined here, but for the purposes of this section, an intuitive understanding will be perfectly enough. At the end of this book, the student will have a much more complete understanding of its meaning.

---

[8] Remember that Supervised learning (SL) is the machine learning task of learning a function that maps an input to an output based on training data that has been labelled.

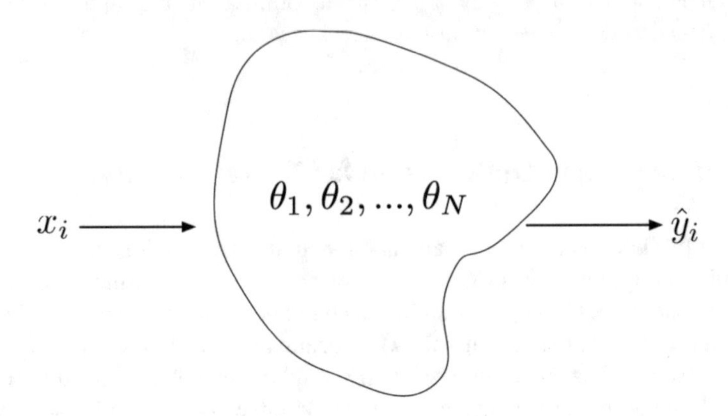

Fig. 3.7 a diagram that gives an intuitive understanding of what a neural network is. x_i are the inputs (for $i = 1, \ldots, n$), θ_i are numbers (or parameters) (for $i = 1, \ldots, N$), and $\hat{y}$ is the output of the network. The network itself is intuitively depicted as the irregular shape in the middle of the figure.

3.8.1 First Definition of Learning

Let us now give a more mathematical formulation of "learning". For simplicity of notation, assume that each input is a monodimensional array $x_i \in \mathbb{R}^n$ with $i = 1, \ldots, M$. In the same way, we will assume that the output will be a monodimensional array $\hat{y}_i \in \mathbb{R}^k$ with k some integer. We will assume that we have a set of M input observations with expected target variables $y_i \in \mathbb{R}^k$. We also assume that we have a mathematical function $L(\hat{y}, y) = L(f(\theta, \hat{y}), y)$ called **loss function**, where we have used the vector notation $y = (y_1, \ldots, y_M)$, $\hat{y} = (\hat{y}_1, \ldots, \hat{y}_M)$ and $\theta = (\theta_1, \ldots, \theta_N)$. This function will be a measure of how "close" expected (y) and predicted ($\hat{y}$) values are, given specific values of the parameters θ_i. We will not define yet how this function may be looking like, as this is not relevant for this discussion yet. Let us summarise the notation we have defined so far.

- $x_i \in \mathbb{R}^n$: input observations (for this discussion we will assume that they are a mono-dimensional array of dimension $n \in \mathbb{N}$). Examples could be age, weight, and height of a person, grey level values of pixels in an image, and so on.
- $y_i \in \mathbb{R}^k$: target variables (which we would like the neural network to predict). Examples could be class of an image, what movie to suggest to a specific viewer, the translated version of a sentence in a different language, and so on.
- $f(\theta, x_i)$: network function. This function will be built with neural networks and will depend on the specific architecture used (feed-forward, convolutional, recurrent, etc.).
- $\theta = (\theta_1, \ldots, \theta_N)$: a set of real numbers, also called parameters or weights.

- $L(\hat{y}, y) = L(f(\theta, x), y)$: loss or cost function. This function is a measure of how "close" y and $\hat{y}$ are. Or, in other words, how good the neural network predictions are.

Those are the fundamental elements that we will need to understand the basics of learning with neural networks.

Info ★ Assumption in the Formulation

For the student who already has some experience with neural networks it is important to discuss one important assumption that has been made silently. Note that skipping this short section in a first reading of this book will not impact the understanding of the rest. If you don't understand the points discussed here, feel free to skip this part and come back later.
The most important assumption here can be found in the way the loss function $L(\hat{y}, y)$ is written. In fact, as written, it is a function of all the M inputs, and this translates in **not** using any mini-batch during the training. The assumption here is that we will measure how good the network's predictions are by considering **all** the inputs and outputs simultaneously. This assumption will be lifted in the following sections. The experienced reader may notice that this will lead to advanced optimisation techniques such as stochastic gradient descent and the concept of mini-batch. Using all the M components of the two vectors $\hat{y}$ and y makes the learning generally slower, although in some situations more stable.

With the notation defined previously, we can now give a formal definition of learning.

Definition 3.5 Given a set of tuples (x_i, y_i) with $i = 1, \dots M$, a mathematical function $f(\theta, \hat{y})$ (the network function) and a function (the loss function) $L(\hat{y}, y) = L(f(\theta, x), y)$ the problem of *learning* is equivalent to minimise the loss function with respect to the parameters θ. Or in mathematical notation finding the θ^*

$$\theta^* = \underset{\theta \in \mathbb{R}^N}{\arg\min}\ L(f(\theta, \hat{y}), y) \tag{3.43}$$

The typically used term for *learning* is training and that is the one we will use in this book. Basically, training a neural network is nothing more than minimising a very complicated function that depends on a very large number of parameters (sometimes billions). This presents very difficult technical and mathematical challenges. For now, this understanding will be sufficient to start understanding how we can tackle this problem.

In what follows, we will discuss how to solve the problem of minimising a function in general and discuss the fundamentals theoretical concepts that are necessary to understand more advanced topics.

3.8.2 Constrained vs. Unconstrained Optimisation

The task of minimising a function, as outlined in the previous section, falls under the category of an unconstrained optimisation problem. This problem can be extended to a more general form by incorporating constraints into the optimisation equation. This can be formulated in the following way: we want to minimise a generic function $g(x)$ subject to a set of constraints

$$\begin{cases} c_i(x) = 0, \ i = 1, \ldots, C_1 \text{ with } C_1 \in \mathbb{N} \\ q_i(x) \geq 0, \ i = 1, \ldots, C_2 \text{ with } C_2 \in \mathbb{N} \end{cases} \tag{3.44}$$

Here $c_i(x)$ and $q_i(x)$ are constraint functions that define some equations and inequalities that need to be satisfied. In the context of neural networks, you may have the constraint that the output (suppose for a moment that $\hat{y}_i$ is simply a number) must lie in the interval $[0, 1]$. Or maybe it must be always greater than zero or smaller than a certain value. Or another typical constraint that we will encounter is when you want the network to output only a finite number of outputs, for example, in a classification problem.

Let us give an example. Let us suppose that we want our network output to be $\hat{y}_i \in [0, 1]$. Our learning problem could be formulated as

$$\min_{\theta \in \mathbb{R}^N} L(f(\theta, x), y) \text{ subject to } f(\theta, x_i) \in [0, 1], \ i = 1, \ldots, M \tag{3.45}$$

Or even more generally

$$\min_{\theta \in \mathbb{R}^N} L(f(\theta, x), y) \ \text{ subject to } f(\theta, x) \in [0, 1] \ \forall \ \theta, x \tag{3.46}$$

This is clearly a *constrained optimization* problem. When dealing with neural networks, this problem is typically reformulated by designing the neural network in such a way that the constraint is automatically satisfied and learning is brought back to an unconstrained optimisation problem.

Info ★ Reducing a Constrained to an Unconstrained Optimisation Problem

The student may be confused by the previous section and wonder how constraints can be integrated into the design of network architecture. This typically occurs in the output layer of the network. For example, in the examples discussed in the previous section, to ensure that $f(\theta, \hat{y}_i) \in [0, 1]$ $i = 1, \ldots, M$ is enough to use the sigmoid function $\sigma(s)$ as activation function for the output neuron. This will guarantee that the network output will always be between 0 and 1 since the sigmoid function maps any real number to the open interval $(0, 1)$. If the output of the neural network should always be 0 or greater, one could use the ReLU activation function for the output neuron.

Building constraint into the network architecture is extremely useful and it typically makes learning much more efficient. Constraints typically come from a deep understanding of the data and the problem you are trying to solve. It pays to find out as many constraints as possible and to try to build them into the network architecture.
Another example of a constrained optimisation problem is when you have a classification problem with k classes. Typically, you want your network to output k real numbers p_i with $i = 1, \dots, k$, where each p_i could be interpreted as the probability of the input observation of being in a specific class. If we want to interpret the p_i as probability the following equation must be satisfied

$$\sum_{i=1}^{k} p_i = 1 \tag{3.47}$$

This is realized by having k neurons in the output layer and use for them the *softmax* activation function. This step reframes the problem into an unconstrained optimization problem since the previous equation will be satisfied by the network architecture. If you want to learn more about the softmax function, you can check Section 5.5 where we discussß its mathematical formulation and its possible use cases.

3.9 Optimization Algorithms

So far, we have discussed the idea that learning is nothing less than minimising a specific function, but we have not discussed the issue of how this "minimising" looks like. This is achieved with what is called an "optimisation algorithm", whose goal is to find the location of the (hopefully) global minimum. Note that we are talking here about numerical algorithms that iteratively search for the minimum of a given function. As we discussed, in any practical applications it is almost always impossible to find this solution analytically (probably one of the only notable exceptions is linear regresssion, as we discuss in Section 4.3.5), and thus we have to rely on numerical routines that find, in a stepwise approach, approximations of the location of the minimum that we hope gets better and better with each iteration.

Practically, all unconstrained minimisation algorithms require the choice of a starting point that we will denote by x_0. In the example of neural networks, this initial point would be the initial values of the weights. Typically, starting from x_0, optimisation algorithms will generate a sequence of iterates $\{x_k\}_{k=0}^{\infty}$ that hopefully will converge toward the global minimum. In all practical applications, only a finite number of terms will be generated, since we cannot generate an infinite number of x_k of course. The sequence will stop when no progress can be made anymore (the value of x_k will not change much more[9]) or a specific solution has been reached with sufficient precision. Usually, the rule to generate a new x_k will use information about the function f to be minimised and one or more previous values (often properly weighted) of x_k. In general, there are two main strategies for optimisation algorithms: line search and trust regions. All neural network optimisers use a line search approach.

3.9.1 Line Search and Trust Region Approaches

In the *line search* approach, the algorithm chooses a direction p_k and searches along this direction for a new value x_{k+1} when trying to minimise a generic function $L(x)$ (recall that x is typically multidimensional and thus talking about *directions* make sense). In general, this approach, once a direction p_k has been chosen, consists in solving

$$\min_{\alpha>0} L(x_k + \alpha p_k) \tag{3.48}$$

for each iteration. In other words, one would need to choose the optimal α along the direction p_k. In general, this cannot be solved exactly, thus in practical application (as we will see later) this approach is used by choosing a fixed α, or by reducing it in a way that is easy to calculate (independently of L). α is what is known as *learning rate* when dealing with neural networks and is one of the most important hyper-parameters (for a discussion of what hyper-parameters are see Chapter 9)[10] when training networks. After deciding on a value for α, the new x_{k+1} is determined using the equation.

$$x_{k+1} = x_k + \alpha p_k \tag{3.49}$$

In the *trust region* approach, the information available on L is used to build a model function m_k (typically quadratic in nature) that approximates f in a sufficiently small region around x_k. This approximation is then used to choose a new x_{k+1}. In this book we will not cover trust region approaches, but the interested reader can find a very complete introduction in *Numerical Optimization, 2nd edition* by J. Nocedal and S.J. Wright, published by Springer.

3.9.2 Steepest Descent

The most obvious, and the most used, search direction for line search methods is the steepest direction $p_k = -\nabla L\,(x_k)$. After all, this is the direction along which the function f decreases more rapidly. To prove it, we can use the Taylor expansion for

[9] Don't be annoyed by the intuitive formulation. We will discuss this later.

[10] A Hyper-parameter is a parameter that does not change during the training and is not related to the training data. Even if the learning rate changes according to some fixed strategy, it is still called a hyper parameter, since it does not change due to training data.

$L(x_k + \alpha p)$ and try to determine in which direction the function decreases the most rapidly.

Info ★ The Taylor Expansion

The Taylor expansion is a mathematical concept that allows us to approximate complex functions using simpler polynomial ones. At the heart of this concept is the idea that near any specific point, most smooth functions can be closely mirrored by a polynomial. To understand this intuitively, imagine you are trying to sketch a curved line without using any curved strokes, but only straight lines. The more lines you use, the closer your sketch will resemble the curve. This is essentially what the Taylor expansion does. It starts with a point on the function, known as the expansion point, and uses the function's derivatives at that point to build up an approximation. The simplest form of Taylor expansion is a linear approximation, where the function is represented by a straight line tangent to the function at the expansion point. This is like using just one straight stroke to approximate a small segment of a curve. As you include more terms from the Taylor expansion, which involve higher-order derivatives, your approximation becomes a polynomial that bends and twists to fit the curve more closely. Each additional term in the expansion adds a new feature to the polynomial, allowing it to mimic the function's behaviour more accurately near the expansion point. The beauty of the Taylor expansion is that it can be used to approximate any smooth function to a high degree of accuracy, as long as you are willing to include enough terms. This makes it an extremely powerful tool in physics, engineering, and other fields where dealing with complex functions directly can be challenging.
For a one-dimensional function $f : \mathbb{R} \to \mathbb{R}$ the Taylor expansion around the point $x = a$ can be written as

$$f(x) = f(a) + f'(a)(x-a) + \frac{f''(a)}{2!}(x-a)^2 + \frac{f'''(a)}{3!}(x-a)^3 + \cdots \quad (3.50)$$

while in a multidimensional case, with $f : \mathbb{R}^n \to \mathbb{R}$ the Taylor expansion around the point $\mathbf{x} = \mathbf{a} \in \mathbb{R}^n$ is

$$f(\mathbf{x}) = f(\mathbf{a}) + \nabla f(\mathbf{a}) \cdot (\mathbf{x} - \mathbf{a}) + \frac{1}{2}(\mathbf{x} - \mathbf{a})^T \cdot H(f(\mathbf{a})) \cdot (\mathbf{x} - \mathbf{a}) + \cdots \quad (3.51)$$

Here, $\nabla f(\mathbf{a})$ is the gradient of f at $\mathbf{a}$, and $H(f(\mathbf{a}))$ is the Hessian matrix of f at $\mathbf{a}$. The formula represents $f(\mathbf{x})$ as an infinite sum that entails the derivatives of the function at the point $\mathbf{a}$. The Hessian is defined by

$$H(f(\mathbf{x})) = \begin{pmatrix} \frac{\partial^2 f}{\partial x_1^2} & \frac{\partial^2 f}{\partial x_1 \partial x_2} & \cdots & \frac{\partial^2 f}{\partial x_1 \partial x_n} \\ \frac{\partial^2 f}{\partial x_2 \partial x_1} & \frac{\partial^2 f}{\partial x_2^2} & \cdots & \frac{\partial^2 f}{\partial x_2 \partial x_n} \\ \vdots & \vdots & \ddots & \vdots \\ \frac{\partial^2 f}{\partial x_n \partial x_1} & \frac{\partial^2 f}{\partial x_n \partial x_2} & \cdots & \frac{\partial^2 f}{\partial x_n^2} \end{pmatrix} \tag{3.52}$$

and is the multidimensional version of the second derivative in Equation 3.50.

We will stop at the first order and write

$$L(x_k + \alpha p) \approx L(x_k) + \alpha p^T \nabla L(x_k) \tag{3.53}$$

assuming that α is small enough. Our question (in which direction the function L decreases more rapidly) can be formulated as solving

$$\min_p L(x_k + \alpha p) \text{ subject to } \|p\| = 1 \tag{3.54}$$

Where $\|p\| = 1$ is the norm of the vector p (or in other words $\|p\| = (p_1^2 + \ldots + p_n^2)$). Note we set the norm equal to 1, since we are only interested in the direction. The magnitude of the movement along that direction will be fine-tuned with the learning rate. Using the Taylor expansion and noting that $L(x_k)$ is a constant, we simply must solve for

$$\min_p p^T \nabla L(x_k) \tag{3.55}$$

Always subject to $\|p\| = 1$. Now, indicating with θ the angle between the direction p and $\nabla L(x_k)$ we can write

$$p^T \nabla L(x_k) = \|p\| \, \|\nabla L(x_k)\| \cos\theta \tag{3.56}$$

And is easy to see that this is minimized when $\cos\theta = -1$, or in other words we simply need tp choose

$$p = -\frac{\nabla L(x_k)}{\|\nabla L(x_k)\|} \tag{3.57}$$

as we claimed at the beginning.

Info Another Explanation of the Steepest Descent Direction

Let $L(\mathbf{x})$ be a differentiable function, where $\mathbf{x}$ is a vector in $\mathbb{R}^n$. We want to prove that the direction in which $L(\mathbf{x})$ varies most rapidly is along its gradient $\nabla L(\mathbf{x})$. Consider a unit vector $\mathbf{u}$ in any direction. The rate of change of $L(\mathbf{x})$ in the direction of $\mathbf{u}$ is given by the directional derivative:

$$D_{\mathbf{u}}L(\mathbf{x}) \equiv \nabla L(\mathbf{x}) \cdot \mathbf{u} \tag{3.58}$$

This definition represents the rate of change of the function L at the point $\mathbf{x}$ as we move infinitesimally in the direction specified by $\mathbf{u}$. If L is differentiable at $\mathbf{x}$, the directional derivative can also be computed using the gradient of L. The dot product can be written as:

$$\nabla L(\mathbf{x}) \cdot \mathbf{u} = \|\nabla L(\mathbf{x})\| \|\mathbf{u}\| \cos\theta = \|\nabla L(\mathbf{x})\| \cos\theta \tag{3.59}$$

since $\|\mathbf{u}\| = 1$ (unit vector). Here, θ is the angle between $\nabla L(\mathbf{x})$ and $\mathbf{u}$. The rate of change $D_{\mathbf{u}}L(\mathbf{x})$ is maximised when $\cos\theta = 1$, which occurs when $\theta = 0$. This means that $\mathbf{u}$ is in the same direction as $\nabla L(\mathbf{x})$. Thus, the function $L(\mathbf{x})$ varies most rapidly in the direction of its gradient $\nabla L(\mathbf{x})$. The steepest descent, which is the direction of the fastest decrease, is therefore in the direction of $-\nabla L(\mathbf{x})$.

To summarise, the steepest descent method is a line search method that searches for a better approximation for the minimum along the direction of minus the gradient of the function for every step. This method is known as **gradient descent algorithm**.

3.9.3 ■ Additional Directions for the Line Search Approach

There are of course other directions that may be used but for neural networks, but those can be neglected (unless you are active in research in optimisation algorithms, in that case you probably don't need to read this book). For the interested reader, here is a list of other *directions* that are sometimes used in optimisation problems. Note that in machine learning, you will not (unless you are active in research) have to use those directions.

- **Newton's Direction:** Employed in Newton's method, this direction uses the inverse of the Hessian matrix of the function at the current point. It provides a second-order approximation, potentially leading to faster convergence compared to gradient descent.
- **Conjugate Gradient Direction:** Used in conjugate gradient methods, especially for large-scale optimisation problems. This direction ensures that search directions are conjugate to each other, leading to efficient exploration of space. See, for example, [5].
- **Quasi-Newton Direction:** In quasi-Newton methods, this direction approximates Newton's direction using gradient information. It is used when computing or inverting the Hessian matrix is computationally expensive.
- **Steepest Descent Direction:** Similar to gradient descent but used in more general settings. It involves moving in the direction where the function decreases most rapidly, which is the negative of the gradient.

- **Momentum-Based Direction:** Incorporates the concept of momentum from physics to accelerate convergence in areas with gentle slopes. This approach combines the gradient descent direction with a fraction of the previous step's direction.
- **Simulated Annealing Direction:** A probabilistic technique that allows for random exploration of the search space. It occasionally accepts moves in non-descent directions to escape local minima. See, for example, [6].
- **Subgradient Direction:** Used when dealing with non-differentiable functions. The subgradient generalises the concept of gradient to functions that are not differentiable at certain points.

This list has the goal of giving you an idea of the enormous number of possibilities and their complexity. In any practical application, you will not have to choose a direction or implement complex algorithms. The algorithms you find in libraries as TensorFlow or pyTorch are already optimised and will do everything for you. But it is important to understand on which algorithm those optimisers are based, and in this way, you can judge the limitations or strength of the various types. For the types without reference in the previous list, a good starting point is always the book by Boyd and Vandenberghe [3].

3.9.4 The Gradient Descent Algorithm

The gradient descent (GD) optimiser finds x_{k+1} using the gradient of the function L according to the formula

$$x_{k+1} = x_k - \alpha \nabla L(x_k) . \tag{3.60}$$

Here, just to remind you, α is the learning rate. The GD algorithm is simply a line search algorithm that searches for better approximations along the steepest descent direction. We can make a simple one-dimensional example ($x \in \mathbb{R}$) and try the algorithm. Let us suppose that we want to minimise the function

$$L(x) = x^2 \tag{3.61}$$

This has a clear minimum at $x = 0$ as this is a simple quadratic form. We can find the minimum position analytically with the first derivative

$$\frac{dL(x)}{dx} = 0 \tag{3.62}$$

this implies that $2x = 0 \rightarrow x = 0$. This is indeed a minimum, since

$$\frac{d^2L(x)}{dx^2} = 2 > 0. \tag{3.63}$$

What does the GD algorithm look like in this case? The algorithm will generate a sequence of x_k by using the formula $x_{k+1} = x_k - 2\alpha x_k$ (remember that we are trying to minimise $f(x) = x^2$). We need of course to choose an initial value x_0 and a learning rate α. For a first try, choose $x_0 = 1$ and $\alpha = 0.1$. The generated x_k sequence can be seen in Table 3.1. From the values is evident how, albeit slowly, the

k	x_k for $x_0 = 1$ and $\alpha = 0.1$
0	1
1	0.8
2	0.64
3	0.512
...	...
40	0.00013
...	...
500	10^{-49}

Table 3.1 The sequence x_k generated for the function $L(x) = x^2$ with the parameters $x_0 = 1$ and $\alpha = 0.1$.

GD algorithm converges toward the right answer $x = 0$. That sounds good, right? What could go wrong? Not everything is so easy, and in GD there is a marvellous hidden complexity. Let us rewrite the formula that in this case is used to generate the sequence x_k:

$$x_{k+1} = x_k - 2\alpha x_k = x_k(1 - 2\alpha) \tag{3.64}$$

Consider for example the value $\alpha = 1$. In this case $x_{k+1} = -x_k$. It is easy to see that this generates an oscillating sequence that never converges. In fact, it is easy to calculate $x_1 = -1$, $x_2 = 1$, $x_3 = -1$ and so on. An oscillating sequence will always be generated for all values of $1 - 2\alpha < 0$, or for $\alpha > \frac{1}{2}$. In Figure 3.8 you can see a plot of the sequence x_k for various values of the parameter α. When α is small, the convergence is very slow (dashed line), and as we discussed, for a value $\alpha > \frac{1}{2}$ (dotted line) an oscillating sequence is generated. It is interesting to note that this oscillating sequence converges quite faster than the others. The value $\alpha = 1$ generates a sequence that does not converge, but what happens for $\alpha > 1$? This is a very interesting case, as it turns out that the sequence diverges (albeit oscillating from positive to negative values). In Figure 3.9 you can see the plot of the sequence for $\alpha = 1.01$.

From a numerical point of view, it is easy to get NaN (if you are using Python) or errors. If you are trying neural networks and you get NaN for your loss function (for example), one possible reason may well be a learning rate that is too big.

Info The Importance of the Learning Rate

The learning rate is possibly one of the most important hyper-parameters that you will have to decide when training neural network. Choose it too small, and the training will be very slow, but choose it too big, and the training

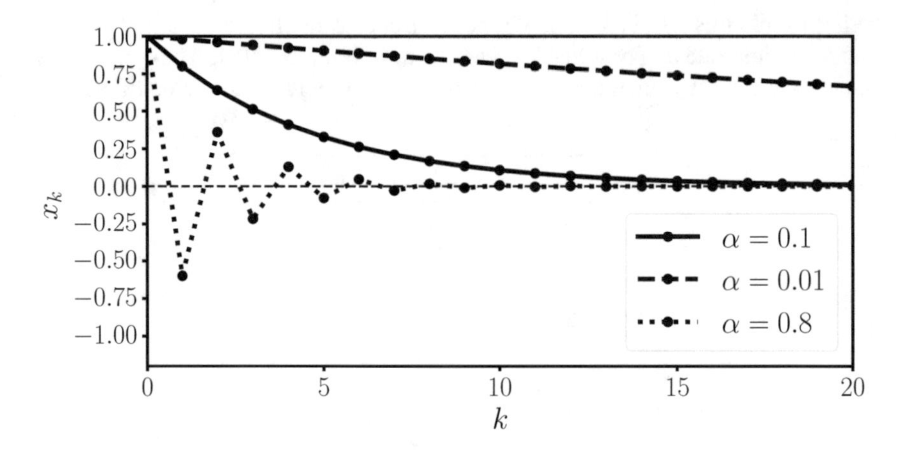

Fig. 3.8 The sequence x_k generated for the function $L(x) = x^2$ for $\alpha = 0.1, 0.01$ and 0.8.

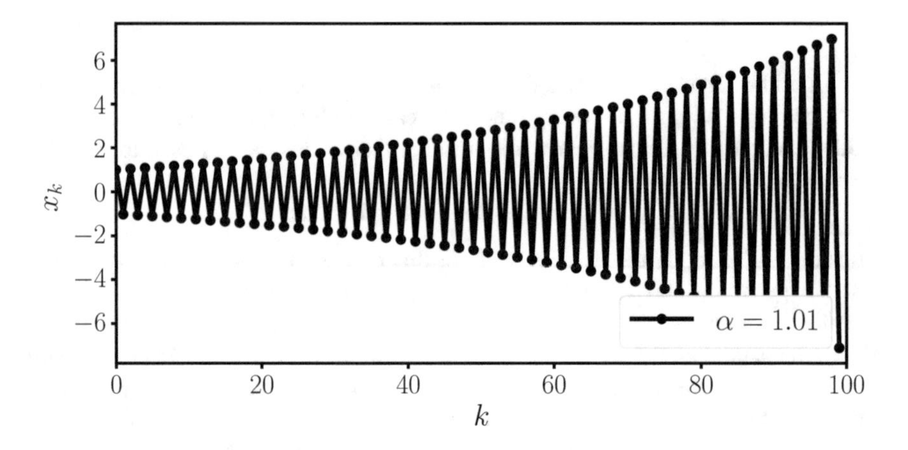

Fig. 3.9 The sequence x_k for the parameter $\alpha = 1.01$. The sequence oscillates from positive to negative values while diverging in absolute value.

will not converge! More advanced optimisers (as Adam for example) try to compensate for this shortcoming by effectively varying the learning rate[a] dynamically, but the initial value is still important.

[a] To be precise, the Adam optimiser does not change the learning rate but uses a different algorithm that is very similar, but not the same, to the gradient descent and therefore sometimes it is said incorrectly that Adam updates the learning rate dynamically.

Now, I must admit that this is a quite trivial case. In fact, the formula for x_k can also be written explicitly as

$$x_k = x_0 (1 - 2\alpha)^k \tag{3.65}$$

And therefore, is easy to see that this sequence converges for $|1 - 2\alpha| < 1$ and diverges for $|1 - 2\alpha| > 1$. For $|1 - 2\alpha| = 1$ it stays at 1 and if $1 - 2\alpha = -1$ it oscillates between 1 and -1. Still, it is quite instructive to see how important the role of the learning rate is when using gradient descent.

3.9.5 Choosing the Right Learning Rate

You may be wondering how to choose the right α at this point. This is a good question, but unfortunately there is no real precise answer, and some clarifications are in order. In all practical cases, you will not use the plain gradient descent algorithm. Consider, for example, that in TensorFlow 2.15 the gradient descent is not even available out of the box, due to its inefficiency. But in general, to check if the (in some cases only initial) learning rate is optimal, follow the steps, assuming that you are trying to minimise a function $f(x)$:

1. You choose an initial learning rate. Typical values[11] are 10^{-2} or 10^{-3}.
2. You let your optimiser run for a certain number of iterations saving each time the $f(x_k)$
3. You plot the sequence $f(x_k)$. This sequence should show convergent behaviour. From the plot you can get an idea if the learning is too small (slow convergence) or too large (divergence). For example, Figure 3.10 shows the sequence $f(x_k)$ for the example we discussed in the previous section. The figure would tell me that using $\alpha = 0.01$ (dashed line) convergence is very slow. Trying larger values for α makes clear how convergence can be faster (continuous and dotted). With $\alpha = 0.1$ after 12-13 iterations you already have a good approximation of the minimum, while for $\alpha = 0.01$ you are still very far.

When training neural networks, always check the behaviour of your loss fFunction. This will give you important information on how the training process is going. This is the reason why when training neural networks, it is important to always check the behaviour[12] of the loss function that you are trying to minimise. Never assume that your model is converging without checking the sequence $L(x_k)$.

3.9.6 Variations of Gradient Descent

To understand variations of GD, the easiest way is to start with the loss function. As we mentioned at the beginning of the chapter, our goal is to minimise the loss

[11] For example, the Adam optimiser in TensorFlow 2.X uses 0.001 as the standard learning rate, unless you specify otherwise.

[12] Tools as TensorBoard (from TensorFlow) have been built with exactly this problem in mind, to give a real time check on how the training is going.

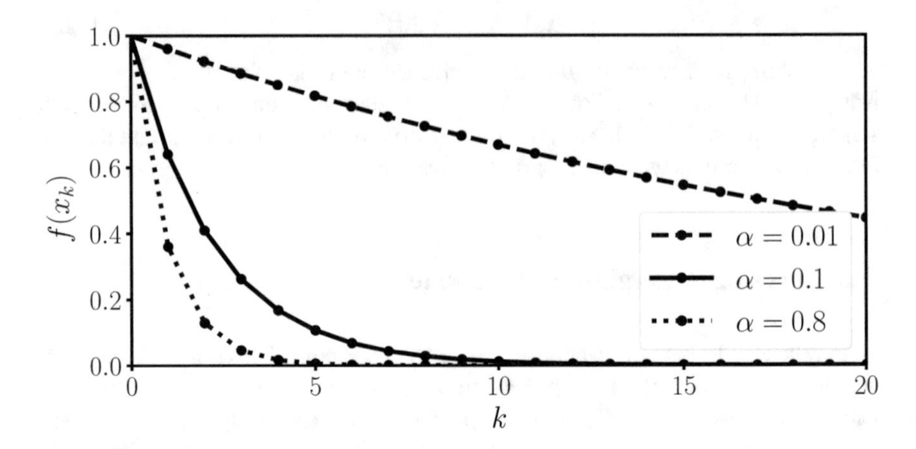

Fig. 3.10 The sequence $f(x_k)$ for the function $f(x) = x^2$ for various values of α.

function $L(f(\theta, x), y)$ where we have used the vector notation $y=(y_1, \ldots, y_M)$, $\hat{y} = (\hat{y}_1, \ldots, \hat{y}_M)$ and $\theta = (\theta_1, \ldots, \theta_N)$. In other words, we have at our disposal M input tuples that we can use. In the plain version of GD, the loss function is written as

$$L(f(\theta, \hat{y}), y) = \frac{1}{M} \sum_{i=1}^{M} l_i(f(\theta, x_i), y_i) \tag{3.66}$$

Where l_i is the loss function evaluated over **one single** observation x_i. For example, we could have a one-dimensional regression problem where our loss function is the mean square error (MSE). In this case, we would have

$$l_i(f(\theta, x_i), y_i) = |f(\theta, \hat{y}_i) - y_i|^2 \tag{3.67}$$

And therefore

$$L(f(\theta, x), y) = \frac{1}{M} \sum_{i=1}^{M} |f(\theta, x_i) - y_i|^2 \tag{3.68}$$

That is the classical formula for the MSE that you may have already seen. In plain GD, we would use this formula to evaluate the gradient that we need to minimize L. Using all M observations have pros and cons.

Info Advantages and Disadvantages of Plain Gradient Descent

Advantages

- Plain GD shows a stable convergence behavior

Disadvantages

- Usually, this algorithm is implemented in such a way that all the dataset must be in memory, therefore it is computationally quite intensive
- This algorithm is typically very slow for very big datasets

Variations of gradient descent are based on the idea of considering only **some** of the observations in the sum in the previous equation instead of all M. The two most important variations are called *mini-batch GD* (MBGD, where you consider a small number of observations $m < M$) and *Stochastic GD* (SGD, where you consider only one observation at a time). Let us look at both in detail starting with the MBGD.

3.9.6.1 Mini-batch Gradient Descent

To clarify the idea behind the method, we can write the loss function as

$$L_m(f(\theta, \hat{y}), y) = \frac{1}{M} \sum_{i=1}^{m} |f(\theta, x_i) - y_i|^2 \qquad (3.69)$$

Where we have introduced $m \in \mathbb{N}$ with $m < M$ called here *batch size*. L_m is defined by summing over m observations sampled from the initial dataset.
Mini-batch GD is implemented according to the following algorithm:

1. A mini-batch size m is chosen. Typical values are 32, 64, 128 or 256 (note that the mini-batch size m does not have to be a power of 2, and could be any number as 137 or 17);
2. $N_b = \lfloor \frac{M}{m} \rfloor + 1$ subsets of observations are created[13] by sampling each time m observations from the initial dataset S without repetition (for a discussion about sampling see Chapter 6). We will indicate them with $S_1, S_2, \ldots, S_{N_b}$. Note that in general if M is not a multiple of m the last batch, S_{N_b} may have a number of observations smaller than m;
3. The parameters θ are updated N_b times using the GD algorithm with the gradient of L_m evaluated over the observations in S_i for $i = 1, \ldots, N_b$;
4. Repeat point 3 until the desired result is achieved (for example the loss function does not vary that much anymore).

When training neural networks, you may have heard the term **epoch** instead of iteration. An epoch is *finished* after all the data have been used in the previous algorithm. Let us give an example. Suppose we have $M = 1000$ and we choose $m = 100$. The parameters θ will be updated using each time 100 input observations. After 10 iterations (M/m) the network will have used all M observations for its training. At this point, it is said that one epoch is finished. An epoch in this example will consist of 10 parameters update (or 10 iterations). Here are advantages and disadvantages.

[13] The symbol $\lfloor x \rfloor$ indicates the integer part of x.

Info Advantages and Disadvantages of Mini-batch Gradient Descent

Advantages

- The model update frequency is higher than with plain gradient descent but lower than SGD (see below), therefore allowing for a more robust convergence than SGD.
- This method is computationally much more efficient than plain gradient descent or Stochastic GD since less calculations (as in SGD) and resources (as in Plain GD) are needed
- This variation is by far (as we will see later) the fastest of the three and the most used

Disadvantages

- The use of this variation introduces a new hyper-parameter that needs to be tuned: the batch size (number of observations in the mini-batch)

We can formally define an **epoch** as

Definition 3.6 An epoch is *finished* after all the input data have been used to update the parameters of the neural network. Recall that in one epoch the parameters of the network may be updated many times.

3.9.6.2 Stochastic Gradient Descent

SGD is also a very commonly used version of GD, and is simply the mini-batch version with $m = 1$. This means updating the parameters of the network by using one observation at a time for the loss function. Of course, this also has advantages and disadvantages.

Info Advantages and Disadvantages of Stochastic Gradient Descent

Advantages

- The frequent updates allow an easy check on how the model learning is going (you don't need to wait until all the dataset has been considered).
- The convergence process is intrinsically noisy, and that may help the model to avoid local minima when trying to find the absolute minimum of the cost function.

Disadvantages

- On large dataset this method is quite slow, since is very computationally intensive due to the continuous updates.

- The fact that the algorithm is noisy can make it difficult for the algorithm to settle on a minimum for the cost function, and the convergence may be not as stable as expected.

3.9.7 How to Choose the Right Mini-batch Size

Now what is the right mini-batch size m? Typical values used by practitioners are of the order of 100 or less. For example, Google TensorFlow standard value (if you don't specify otherwise) is 32. Why this value? What is so special? To understand the reasons, we need to study the behaviour of MBGD for various choices of m. To make it resemble real cases, we will use the Zalando dataset [7]. You may have already seen it. It is a dataset that contains 70000 images of 10 types of clothing. The images are gray-level 28x28 pixel images. You can see three examples from the dataset in Figure 3.11. We will build a classifier[14] and check its convergence behaviour. I have

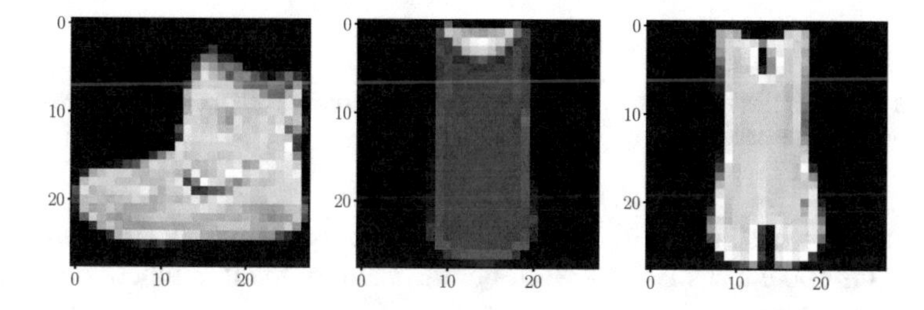

Fig. 3.11 Three images from the Zalando dataset (© Zalando SE, 2017). They are 28×28 pixel, gray level images. The dataset contains 70000 images of 10 types of clothing items.

trained the network for 10 epochs (remember what an epoch is?) on 60000 training images and I have measured the running time[15] needed, the reached value of the loss function and the accuracy at the end of the training. I have used the following values for the mini-batch size m: 5, 10, 100, 1000, 3000, 6000 and 60000 (effectively using all the training data, so no mini-batch). Note that while for $m = 10$ the required time

[14] For those of you interested, the results have been obtained with a neural network with two layers each having 128 neurons with the ReLu activation function, by using the Adam optimizer. First, if you don't exactly know what I am talking about, you can simply skip those details. The discussion can be followed even without understanding the details of how the network is designed. Secondly, using Adam is only for practical reasons, as in TensorFlow the MBGD is not even available out of the box. But the conclusions continue to be valid.

[15] I have run these tests on a modern macbook pro from 2021, and that means an 2.3 GHz 8-Core Intel Core i9 without GPU.

is 2.34 min, when using $m = 1$ 19.18 minutes are needed for 10 epochs (this on a macbook pro from 2020)! In Figure 3.12 you can see the results of this study.

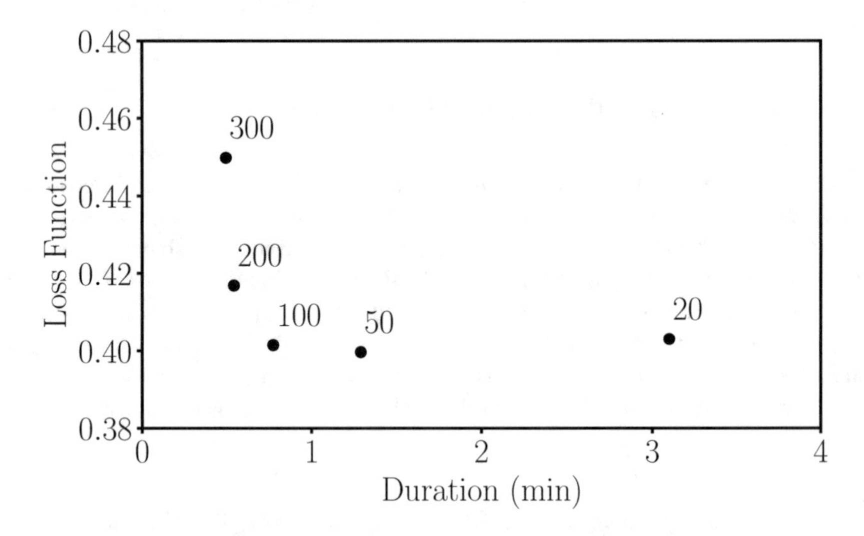

Fig. 3.12 A plot of the loss function value reached after 100 epochs on the Zalando dataset plotted vs. the running time needed.

Let us see what Figure 3.12 tells us. When we use $m = 300$, the running time required for 100 epochs is the lowest, but the value of the loss function reached is higher. Decreasing m decreases the reached loss function value quite rapidly until we reach the "elbow". Between $m = 200$ and $m = 50$ the behaviour changes. Decreasing m does not decrease the value of the loss function much, but the running time becomes longer and longer. So, when we reach the elbow decreasing m does not bring many advantages anymore. As you will notice, around the elbow m is of the order of 100. Figure 3.12 is an explanation for why typical values for m are of the order of 100. Of course, the optimal value is dependent on the data, and some testing is required, but in most cases a value around 100 or 50 is a very good starting point.

3.9.8 ★ Stochastic Gradient Descent and Fractals

We have discussed in previous sections how choosing the wrong learning rate can slow down convergence or even make it diverge. But the discussion done was for a one-dimensional case and thus was very simple. In this section, I want to show you how much complexity is hidden when using SGD. I want you to see how specific ranges of the learning rate make the convergence chaotic (in the mathematical sense of the word), thus showing one of the many hidden gems that you can find when dealing with optimisation problems. Let us consider the problem[16] in which our M inputs $x^{[i]}$are bi-dimensional, in other words $x^{[i]} = (x_1^{[i]}, x_2^{[i]}) \in \mathbb{R}^2$. We call our target variables $y^{[i]}$. The optimisation problem we are trying to solve involves minimising the function.

$$L = \frac{1}{2} \sum_{i=1}^{M} \left[f(x_1^{[i]}, x_2^{[i]}) - y^{[i]} \right]^2 \tag{3.70}$$

with

$$f(x_1^{[i]}, x_2^{[i]}) = w_1 x_1^{[i]} + w_2 x_2^{[i]} \tag{3.71}$$

A simple linear combination of the inputs. The problem is simple enough. We minimise the MSE (Mean Square Error) and try to find the best parameters w_1 and w_2 that minimise L. Let us simply further illustrate the problem. Consider $M = 3$ inputs. In particular, to make it more concrete, consider the input matrix[17]

$$X = \begin{pmatrix} 0 & 1 \\ 1 & \frac{1}{2} \\ 1 & -\frac{1}{2} \end{pmatrix} \tag{3.72}$$

We write our labels also in matrix form

$$Y = \begin{pmatrix} 0 \\ 4 \\ 0 \end{pmatrix} \tag{3.73}$$

Note that what I will show you here is not dependent on the numerical values. You can reproduce the results with different values without problems. Let us first find exactly the minimum of L (since in this easy case we can do that). To do that, we need simply to derive L and solve the two equations

$$\frac{\partial L}{\partial w_1} = 0; \frac{\partial L}{\partial w_2} = 0 \tag{3.74}$$

Calculations are boring but not overly complex. By solving the two above-mentioned equations, you will find that the minimum is at $x^* = (2, 4/3)$. To implement a SGD optimiser, the following algorithm can be followed:

1. Choose a learning rate α;
2. Choose a random value between from $\{1,2,3\}$ and assign it to i
3. Update the parameters w_1, w_2 by using

---

[16] This problem is an adaptation of the one described in Rojas, R. (2013). Neural networks: a systematic introduction. Springer Science & Business Media.

[17] Note that all the inputs can be written in matrix form for simplicity.

$$l_i = \frac{1}{2}\left(f\left(x_1^{[i]}, x_2^{[i]}\right) - y^{[i]}\right)^2 \tag{3.75}$$

in other words we use l_i to calculate the derivatives to update the weights according to the Gradient Descent rule $w_j \to w_j - \alpha \partial l_i / \partial w_j$ for $j = 1, 2$; each time save the values w_1, w_2, for example in a python list;

4. Repeat points 2 and 3 a certain number of times N;

By following the previous algorithm, we can plot in the (w_1, w_2) space all the points we have obtained and saved in point 3 above. These are all the values that the two parameters w_1 and w_2 will assume during the optimisation procedure. In Figure 3.13 you can see the result for $\gamma = 0.65$. The result is nothing short of amazing.

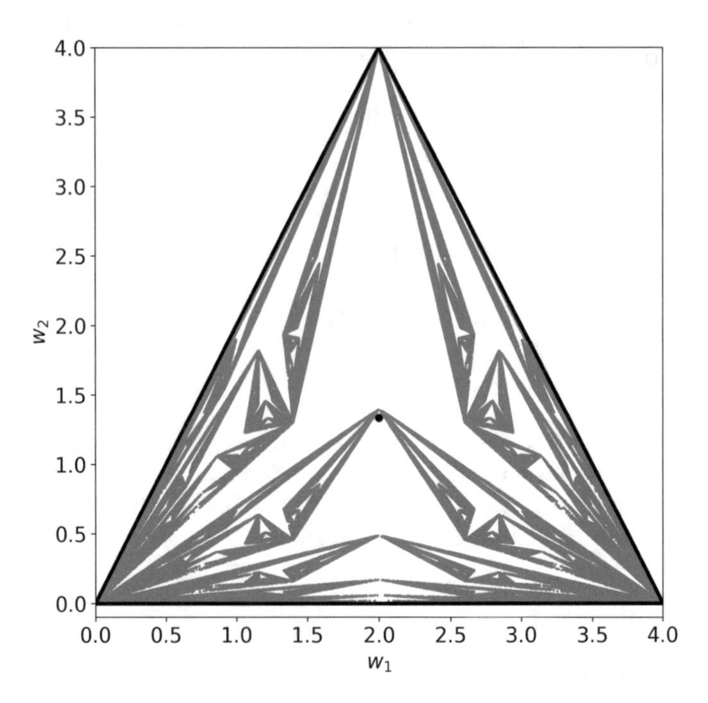

Fig. 3.13 Each blue point is a tuple of values (w_1, w_2) that are generated by using SGD as described in the section for a value of the learning rate $\gamma = 0.65$. The plot has been obtained with $4 \cdot 10^5$ iterations. The black point at the center of the figure is the minimum of the loss function $x^* = (2, 4/3)$.

It can be shown that what you see in Figure 3.13 is indeed a fractal. The mathematical proof is way beyond the scope of this book, but in case you are interested, you can consult the beautiful book *Fractals Everywhere*, by M.F. Barnsley published by Dover. One of the main properties of fractals is that when you zoom in a detail, you will find the same structure that you observe at a larger scale. To convince you, at least intuitively, that this is what is happening, Figure 3.14 shows you a detail of Figure 3.13. In the zoomed area you can observe the same kind of structure that you see at a larger scale.

The particular structure of the fractal depends on the learning rate. In Figure 3.15 you can see the fractal structure for different learning rates, from 0.65 to 1.0. It is quite fascinating to see how the structure changes, showing the great complexity that is hidden in the use of SGD.

Now when using smaller learning rates, at a certain point the fractal structure completely disappears quite suddenly, leaving an unstructured cloud of points, as you can see in Figure 3.16. The smaller the learning rate, the smaller the cloud of points.

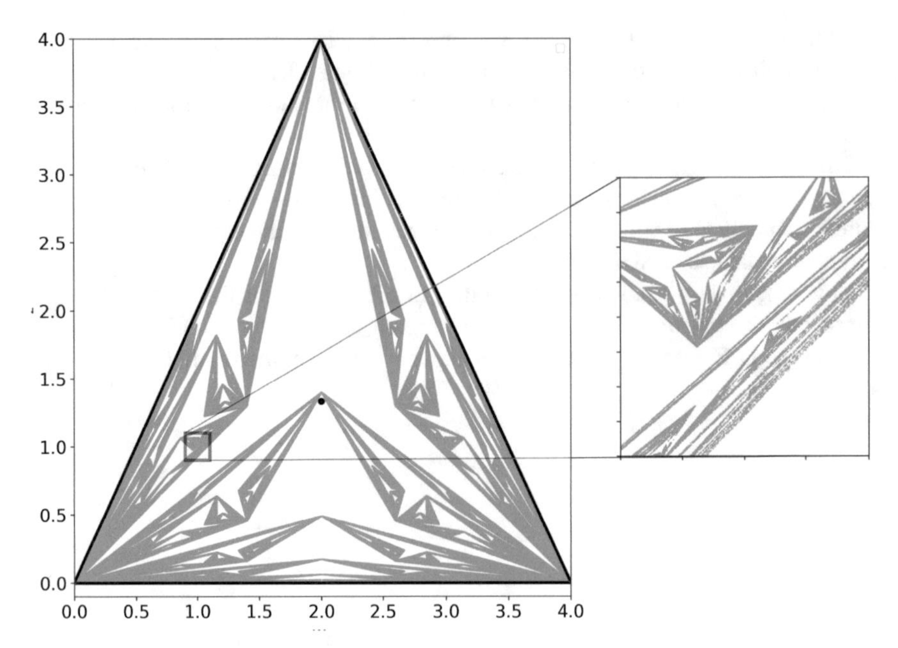

Fig. 3.14 A zoomed region that shows the fractal nature that the SGD algorithm can generate. This picture has been generated with a learning rate of $\gamma = 0.65$ and with 10^7 iterations. In the zoomed region you can clearly see the same kind of structure that you observe at a larger scale on the left. The zoomed region is less sharp than the one on the left since only a fraction of the 10^7 points happen to be in the small zoomed in region.

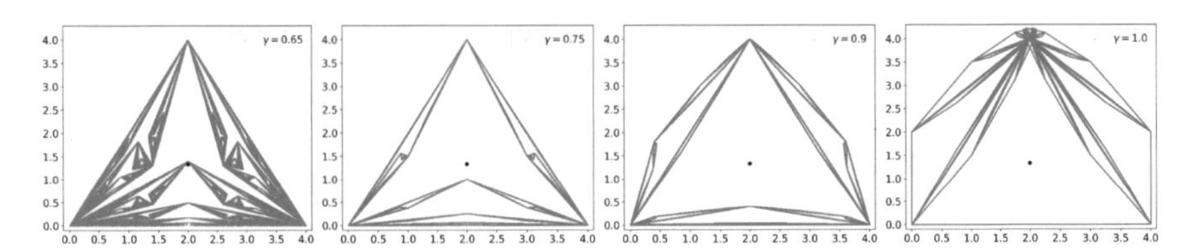

Fig. 3.15 fractal shapes obtained by SGD for different learning rates.

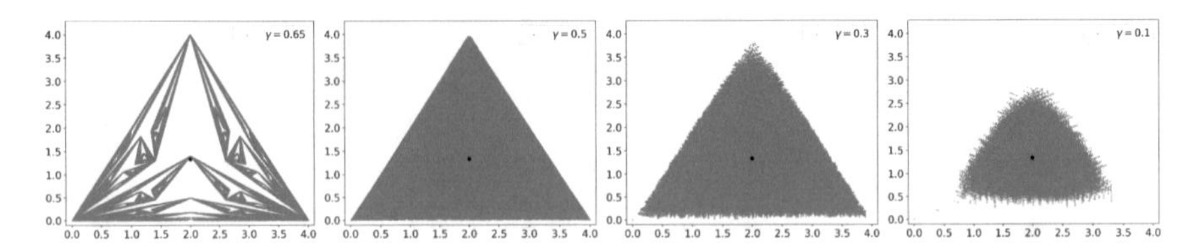

Fig. 3.16 Choosing smaller and smaller learning rates fractal structures completely disappear, leaving an unstructured cloud of points centred on the global minimum x^* of L.

Finally, by choosing a very small learning rate, for example $\gamma = 5 \cdot 10^{-4}$, SGD delivers the behaviour we would expect: the algorithm converges and remains close to the expected minimum. You can see how the plot looks in Figure 3.17. the Gradient

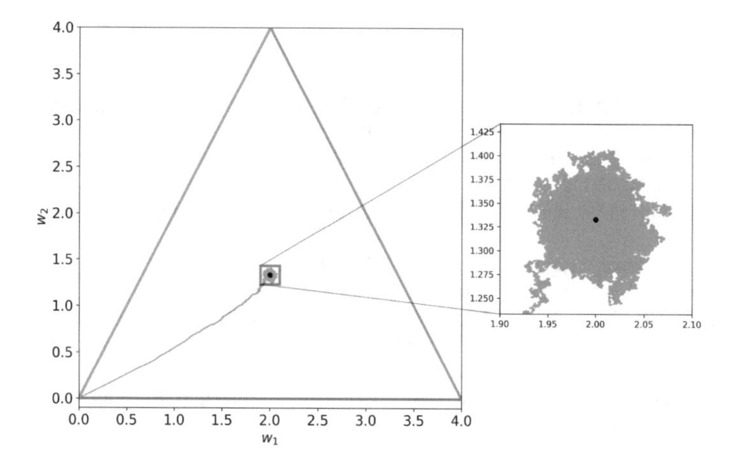

Fig. 3.17 by choosing a very small learning rate $\gamma = 5 \cdot 10^{-4}$ SGD will move in the direction of the expected global minimum x^* and remain in its vicinity, as can be seen from the zoomed-in region. However, SGD continues to deliver points that remain around x^*, but never converge to it. The lower the learning rate, the smaller the cloud of points around x^*.

Descent algorithm, especially in its Stochastic version, has an incredible hidden complexity, even for a trivial case, as the one described in the previous section. This is the reason why training neural networks can be so difficult and tricky and why choosing the right learning rate and optimiser is so important.

3.10 Conclusions

After having read this chapter, you will appreciate more why calculus is so important for machine learning. It offers the mathematical framework required for understanding optimisers (algorithms that effectively train machine learning models). At its core, machine learning is about making predictions and decisions based on data, which involves adjusting the algorithm parameters to minimise the error between expected and predicted output. This is achieved by using derivatives in various approaches.

Furthermore, calculus is indispensable for understanding the behaviour of algorithms as they learn from data. For instance, in neural networks, calculus is at the basis of the backpropagation process, where the derivative of the loss function is computed with respect to each weight. This process relies on the chain rule of calculus to efficiently compute gradients for each layer of the network, enabling the model to learn complex patterns from input data.

References

1. David E Rumelhart, Geoffrey E Hinton, and Ronald J Williams. Learning representations by back-propagating errors. *Nature*, 323(6088):533–536, 1986.
2. Louis Kirsch and Jürgen Schmidhuber. Meta Learning Backpropagation And Improving It, March 2022. arXiv:2012.14905 [cs, stat].
3. Stephen P Boyd and Lieven Vandenberghe. *Convex optimization*. Cambridge university press, 2004.
4. Jorge Nocedal and Stephen J Wright. *Numerical optimization*. Springer, 1999.
5. Magnus R. Hestenes and Eduard Stiefel. Methods of conjugate gradients for solving linear systems. *Journal of Research of the National Bureau of Standards*, 49(6):409–436, 1952.
6. S. Kirkpatrick, C. D. Gelatt, and M. P. Vecchi. Optimization by simulated annealing. In *Science*, volume 220, pages 671–680, 1983.
7. Han Xiao, Kashif Rasul, and Roland Vollgraf. Fashion-mnist: a novel image dataset for benchmarking machine learning algorithms, 2017.

Chapter 4
Linear Algebra

'Obvious' is the most dangerous word in mathematics.
– Eric Temple Bell, Scottish mathematician.

This chapter provides an essential introduction to linear algebra, tailored to improve understanding of its importance in machine learning. It begins by elucidating the fundamental concepts of vectors and matrices, essential building blocks, and delves into their various operations such as addition, subtraction, multiplication, and more advanced procedures such as dot and cross products. The discussion then extends to the critical notions of eigenvectors and eigenvalues, highlighting their significance in matrix analysis and optimisation problems prevalent in machine learning. Central to the chapter is the exploration of Principal Component Analysis (PCA), a key technique in dimensionality reduction, which uses eigenvectors and eigenvalues for dimensionality reduction. The chapter emphasises the practical application of these linear algebraic concepts in deep learning, illustrating their indispensability in algorithm design and model optimisation.

4.1 Motivation

Linear algebra, in probably the most simplified description ever conceived by anyone, concerns itself with arrays of numbers called vectors and matrices. These data structures appear continuously when dealing with any kind of data you will encounter in machine learning. Tables of data where each row makes up one observation and each column is a feature describing the data are the most common type of data. If you have difficulty imagining what I am talking about, imagine a Microsoft Excel sheet full of numbers. Practically any type of data can be presented in an array form, even what is typically called *unstructured*. Images are 3-dimensional arrays made up of their colour values expressed as three numbers (red, green, and blue values) at each x and y positions. In Figure 4.1 you can see that an image is nothing more than an array of numbers. In the left panel, you can see the famous portrait of Einstein. The middle panel is the same image reshaped to 20×27 pixels with the grey value of each pixel overimposed. In the right panel, you see only the array of numbers that make up the image. Sound is similar and can be converted analogously into an array

U. Michelucci, *Fundamental Mathematical Concepts for Machine Learning in Science*,
https://doi.org/10.1007/978-3-031-56431-4_4

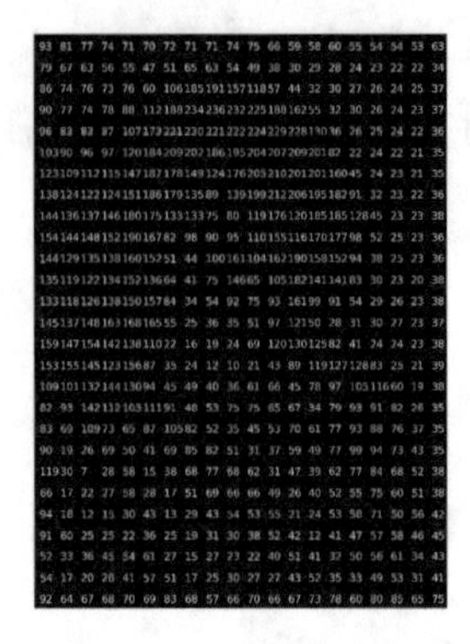

Fig. 4.1 Here you can see how an image is nothing else that an array of numbers. In the left panel, you can see the famous portrait of Einstein. The middle panel is the same image reshaped to 20×27 pixel with the grey value of each pixel overimposed. In the right panel you see only the array of numbers that make up the image.

of numbers.

It is therefore natural to expect that in machine learning arrays of numbers (in particular matrices) play a crucial role. Understanding linear algebra is therefore fundamental to be able to understand how machine learning works and to implement the algorithms efficiently.

In machine learning, almost everything is a matrix[1] (the weights of a neural network, the input dataset, etc.). All Python libraries that are used to implement and train neural networks today use multidimensional arrays as data types (the two most important libraries are pyTorch [1] and TensorFlow [2]). Arrays of numbers are so important that the name of the TensorFlow library from Google comes from *Tensor*, that is, in its most simple definition, a multi-dimensional array of numbers.

Info Vectorisation in Python

Linear algebra is such an important aspect of machine learning, that entire Python libraries are built around it. `numpy` for example is the most known Python numerical library. At its core lie functions implemented in C++ that take advantage of arrays and allow to perform several important operations in deep learning in one shot instead of expensive (from a computational point of view) loops.
For example, suppose that you want to multiply two arrays of 10^7 numbers element-wise. In Python you can easily implement it as a loop or you can use the specifically optimised `numpy` function `multiply()`. This second option is roughly 100 times faster than a simple loop. In machine learning, you

[1] To be technically precise, often a multi-dimensional array. Recall that a matrix is a 2-dimensional array of numbers; multidimensional arrays of numbers are called tensors.

perform operations over and over again, thus making this kind of optimisation extremely important.

This chapter contains the most basic concepts that anyone who wants to use deep learning should understand. Many topics in linear algebra are very interesting and important but could not be included in this short chapter for space reasons (this book is not a course on linear algebra) and most importantly because they are not used as often in machine learning. The reader interested in learning more can consult several books, such as, for example, [3, 4, 5].

4.2 Vectors

Let us start with the simplest form of array: a one-dimensional list of numbers called a *vector*. Vectors are ordered collections of numbers (real or complex). They typically are indicated with a letter in boldface, with a line below the symbol, or an arrow above. If a vector contains real numbers, it is called a *real* vector, if it contains complex numbers, it is called a *complex* vector. In this book, we will deal only with real vectors. A vector are indicated by the following symbols

$$\mathbf{x},\ \underline{x},\ \overrightarrow{x} \tag{4.1}$$

In this book we will denote vectors with a bold-face letter, e.g. $\mathbf{x}$. Some examples of vectors are

$$\begin{aligned}
\mathbf{x} &= (2, 3, 6) \\
\mathbf{x} &= (1, 2, 3.4, 6.4, 6) \\
\mathbf{x} &= (\pi, \pi/2, 1) \\
\mathbf{x} &= (2, 3, 6).
\end{aligned} \tag{4.2}$$

In general, the components of a n-dimensional vector $\mathbf{x} \in \mathbb{R}^n$ are indicated with x_i with $i = 1, ..., n$. So we can write

$$\mathbf{x} = (x_1, x_2, ..., x_n). \tag{4.3}$$

Sometime in books and on websites, vectors are represented in horizontal notation (as we have done so far) and some time in vertical. For example you can find

$$\mathbf{x} = (x_1, x_2, ..., x_n) \tag{4.4}$$

or

$$\mathbf{x} = \begin{pmatrix} x_1 \\ x_2 \\ \vdots \\ x_n \end{pmatrix} \tag{4.5}$$

Technically speaking, this should not be relevant if the operations that you want to perform on vectors are correctly defined. Additionally, as long as one deals only with vectors, this distinction, horizontal and vertical, is completely irrelevant. We will discuss this again when we combine vectors and matrices.

When we sum or subtract vectors, this is done element-wise. For example, consider two vectors $\mathbf{x} \in \mathbb{R}^n$ and $\mathbf{y} \in \mathbb{R}^n$. We have

$$\mathbf{x} + \mathbf{y} = (x_1 + y_1, x_2 + y_2, ..., x_n + y_n) \tag{4.6}$$

Note that we can sum only vectors that have the same number of components. In the same way, we will have

$$\mathbf{x} - \mathbf{y} = (x_1 - y_1, x_2 - y_2, ..., x_n - y_n) \tag{4.7}$$

Multiplication between vectors is not so easily defined. We will deal with it in the next section. Division between vectors cannot be uniquely defined, as will be discussed at the end of Section 4.2.3.

Info Element-wise Sum in Python

When you are programming in Python and you use, for example, `numpy` [6] or `TensorFlow` [2], you will realise very soon that the standard operations between numpy or TensorFlow arrays are done element-wise. For example, if `a` and `b` are two numpy arrays, `a+b` is the sum of the two arrays done element-wise, as you would expect from the previous discussion. Things get slightly more complicated when you multiply arrays, as we will discuss in Section 4.2.3.

4.2.1 Geometrical Interpretation of Vectors

We have defined the vector in an abstract way as an ordered collection of numbers. But it is easy to get an intuitive understanding of what a vector is if we restrict ourselves to 2 or 3 dimensions.

For example, in two dimensions, a vector $\mathbf{x} = (x_1, x_2)$ can be thought of as an arrow in an x, y space with two components x_1 and x_2 along the x and y axis, respectively. The length of the arrow is typically indicated with $|\mathbf{x}|$ and is called its norm (see Section 4.2.2 for more details). In Figure 4.2 you can see a graphical representation of a vector in two dimensions.

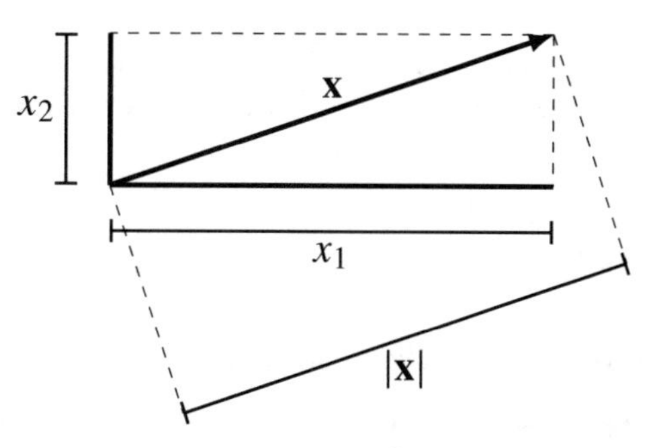

Fig. 4.2 A geometrical interpretation of a vector $\mathbf{x} \in \mathbb{R}^2$. The vector can be thought as an arrow in an x, y space with two components x_1 and x_2 along the x and y axis respectively. The length of the arrow is typically indicated with $|\mathbf{x}|$ and is called its norm (see Section 4.2.2 for more details).

In the geometrical interpretation it is said a vector has a magnitude (the norm) and a direction (related to the sign of the two components). It is understood that if, for example, $x_1 < 0$ and $x_2 < 0$, then the arrow in Figure 4.2 would point in the opposite direction.

The same interpretation can be given in three dimensions. It is better understood graphically from Figure 4.3. Of course if the vector has n dimensions (with $n > 3$)

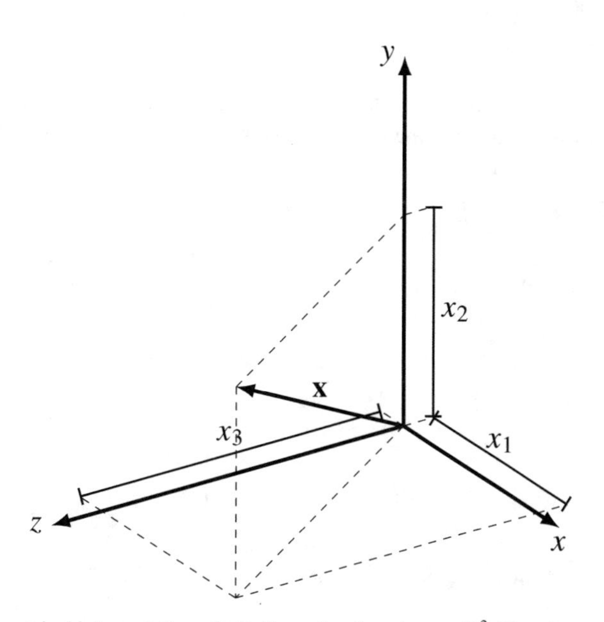

Fig. 4.3 A geometrical interpretation of a 3-dimensional vector $\mathbf{x} \in \mathbb{R}^3$. The vector can be thought as an arrow in an x, y, z space with 3 components x_1, x_2 and x_3 along the x,y and z axis respectively. The length of the arrow is typically indicated with $|\mathbf{x}|$ and is called its norm (see Section 4.2.2 for more details).

then it is impossible to visualise it, but the interpretation remains valid. x_i is the component of the vector along the ith axis in an n-dimensional space.

4.2.2 Norm of Vectors

In general terms, a *norm* can be defined as follows:

Definition 4.1 (Norm Function) A norm is a real-valued function $|\cdot| : \mathbb{R}^n \to \mathbb{R}$ that satisfy the following properties (expressed here in terms of a generic vector $\mathbf{x} \in \mathbb{R}^n$)

1. $|\mathbf{x}| > 0$ for $\mathbf{x} \neq \mathbf{0}$; $|\mathbf{x}| = \mathbf{0} \iff \mathbf{x} = \mathbf{0}$.
2. $|k\mathbf{x}| = |k||\mathbf{x}| \;\; \forall k \in \mathbb{R}$.
3. $|\mathbf{x} + \mathbf{y}| \leq |\mathbf{x}| + |\mathbf{y}|$

Where we have used the notation $\mathbf{0} = (0, 0, \ldots.0)$. There are several norms that are commonly used. Below is a list of the three most common.

1. (p-norm) $|\mathbf{x}|_p = \left(\sum_{i=1}^{n} |x_i|^p \right)^{1/p}$
2. (ℓ_2 norm) $|\mathbf{x}|_2 = \sqrt{x_1^2 + x_2^2 + \ldots + x_n^2}$
3. $|\mathbf{x}|_\infty = \max_{i \in \{1,\ldots,n\}} |x_i|$

It is interesting to note that the norm number 3 is just the limit of the p-norm for $p \to \infty$. The most used and known norm is the ℓ_2 norm, also known as the Euclidean norm.

Info ★ Proof that $|\mathbf{x}|_\infty = \lim_{p\to\infty} |\mathbf{x}|_p$

Theorem 4.1 $\lim_{p\to\infty} |\mathbf{x}|_p = \max_{i\in\{1,\ldots,n\}} |x_i|$

Proof Let us assume that $|x_j|$ is the largest of all the absolute values of the components of $\mathbf{x}$. We can re-write $|\mathbf{x}|_p$ as

$$|\mathbf{x}|_p = \left[|x_j|^p \left(\sum_{i=1, j\neq j}^{n} \left| \frac{x_i}{x_j} \right|^p + 1 \right) \right]^{1/p} = |x_j| \left[\sum_{i=1, j\neq j}^{n} \left| \frac{x_i}{x_j} \right|^p + 1 \right]^{1/p} \tag{4.8}$$

now note that since $|x_j|$ is the largest of the absolute value of all the components we have

$$\left| \frac{x_i}{x_j} \right| < 1 \;\; \forall i \neq j \tag{4.9}$$

and therefore we have

$$\lim_{p\to\infty}\left|\frac{x_i}{x_j}\right|^p = 0 \ \forall i \neq j \tag{4.10}$$

thus by taking the limit of Equation 4.8 for $p \to \infty$ and by using Equation 4.10 we get

$$\lim_{p\to\infty}|\mathbf{x}|_p = |x_j| \lim_{p\to\infty}\left[\sum_{i=1, j\neq j}^{n}\left|\frac{x_i}{x_j}\right|^p + 1\right]^{1/p} = |x_j| \tag{4.11}$$

and that concludes our proof. Just remember that $|x_j|$ is the largest component in absolute value of $\mathbf{x}$. □

If you remember the geometrical interpretation of vectors, it is easy to see how the ℓ_2 norm is the length of the vector in an n-dimensional Euclidean space.

Info ★ Norms in Deep Learning

Norms are widely used in machine learning. For example ℓ_1 and ℓ_2 norms are used in regularisation and many loss functions can be written as some norm of some quantities. For example, indicating with y_i some target variable that we want to predict, and with $\hat{y}_i$ the prediction of some model, the Mean Squared Error (MSE) is given by the equation (for $i = 1, ..., M$)

$$\text{MSE} = \frac{1}{M}\sum_{i=1}^{M}(y_i - \hat{y}_i)^2 \tag{4.12}$$

and can be written with the ℓ_2 norm as

$$\text{MSE} = \frac{1}{M}|\mathbf{y} - \hat{\mathbf{y}}|_2 \tag{4.13}$$

where we have indicated $\mathbf{y} = (y_1, ...y_M)$ and $\hat{\mathbf{y}} = (\hat{y}_1, ..., \hat{y}_M)$.

4.2.3 Dot Product

Multiplying vectors is not as easily defined as summing them. What should the result be from a hypothetical multiplication? A vector? Or maybe a single real (or complex) number? In algebra there are two multiplication operations that give, respectively, a single number and a vector as results.

The first operation we will discuss is called the *dot*-product. The dot product between two vectors $\mathbf{x}$ and $\mathbf{y}$ is indicated by $\mathbf{x} \cdot \mathbf{y}$ and is defined by

$$\mathbf{x} \cdot \mathbf{y} = \sum_{i=1}^{n} x_i y_i \tag{4.14}$$

and again, note that this operation is only possible between vectors of the same length. The results of the dot-product is a single real number[2].

Info Multiplication of Vectors in numpy

In Python you can easily multiply numbers with `numpy`. But you have to be careful. The `multiply()` function multiply two arrays element-wise, but the output will not be a single number, but an array. If you consider two arrays $\mathbf{x} = (x_1, x_2, ..., x_n)$ and $\mathbf{y} = (y_1, y_2, ..., y_n)$, the result of $\texttt{multiply}(\mathbf{x}, \mathbf{y})$ will be an array with components

$$\texttt{multiply}(\mathbf{x}, \mathbf{y}) = (x_1 y_1, x_2 y_2, ..., x_n y_n) \tag{4.15}$$

If you want the dot product as we describe you have to use the `dot()` function in `numpy`. Just keep this in mind and be aware of the difference.

It can be shown that the dot product can be expressed in terms of the ℓ_2 norm of the vectors and the angle between them

$$\mathbf{x} \cdot \mathbf{y} = |\mathbf{x}|_2 |\mathbf{y}|_2 \cos\theta \tag{4.16}$$

So $\mathbf{x} \cdot \mathbf{y}$ is nothing else than the length of one vector multiplied by the length of the projected second vector along the first. To prove this, one can perform, with the help of Figure 4.4, the following calculation

$$\begin{aligned}\mathbf{x} \cdot \mathbf{y} &= x_1 y_1 + x_2 y_2 = |\mathbf{x}|_2 \cos\alpha |\mathbf{y}|_2 \cos(\alpha+\theta) + |\mathbf{x}|_2 \sin\alpha |\mathbf{y}|_2 \sin(\alpha+\theta) = \\ &= |\mathbf{x}|_2 |\mathbf{y}|_2 (\cos\alpha\cos(\alpha+\theta) + \sin\alpha\sin(\alpha+\theta)) = \\ &= |\mathbf{x}|_2 |\mathbf{y}|_2 \cos\theta\end{aligned} \tag{4.17}$$

where in the last passage we have used the formula

$$\cos(\alpha - \theta) = \cos\alpha\cos\theta + \sin\alpha\sin\theta \tag{4.18}$$

Of course the proof would be slightly easier if one would choose the x axis along the $\mathbf{x}$ vector, thus having $\alpha = 0$. However, the proof given is general and is worth understanding for the sake of generality.

[2] If the vectors have complex component, the dot-product will be a complex number naturally.

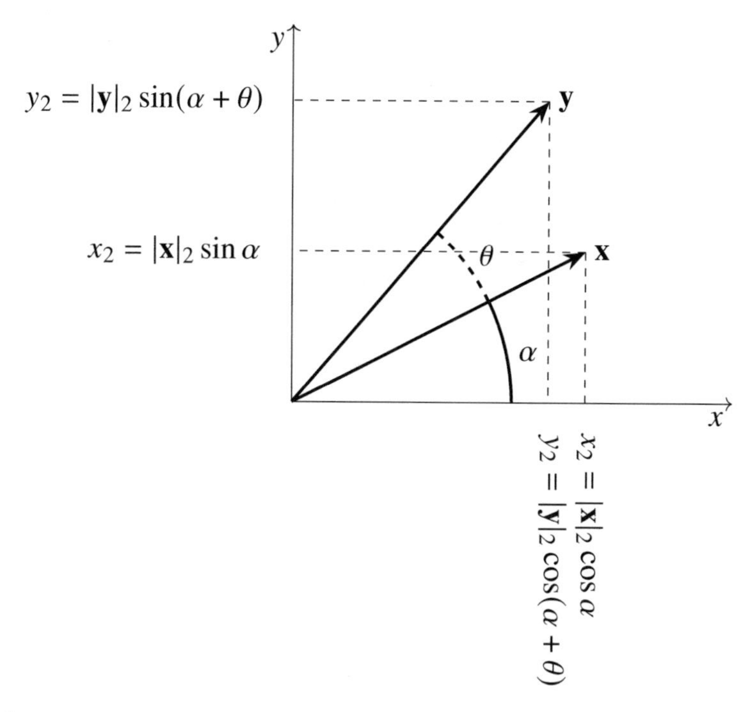

Fig. 4.4 The components along two arbitrary axis can be expressed in terms of the ℓ_2 norm of the vectors multiplied by the cosine and sine of the respective angles as depicted in the figure.

4.2.4 ★ Cross Product

In physics and mathematics there is another operation (called the cross product) that is often used. We report it for completeness, but it is never used in machine learning, so feel free to skip this section if you are not interested. The cross product between two vectors **x** and **y** is indicated with the symbol $\times$ and is defined by

$$\mathbf{x} \times \mathbf{y} = |\mathbf{x}|_2 |\mathbf{y}|_2 \sin(\theta)\mathbf{n} \tag{4.19}$$

where **n** is the unit vector (with ℓ_2 norm equal to one) perpendicular to the plane identified by **x** and **y** in the direction given by the right-hand rule depicted in Figure 4.5. If the vectors **x** and **y** are parallel the cross product is the zero vector **0**. We can express the cross product in terms of the components x_i and y_i **x** and **y** along the axes x, y and z. Indicating with **i**, **j** and **z** the unit vectors along the x, y and z axis respectively[3] it is easy to see that

[3] Assuming that the three axis x, y and z are normal to each other.

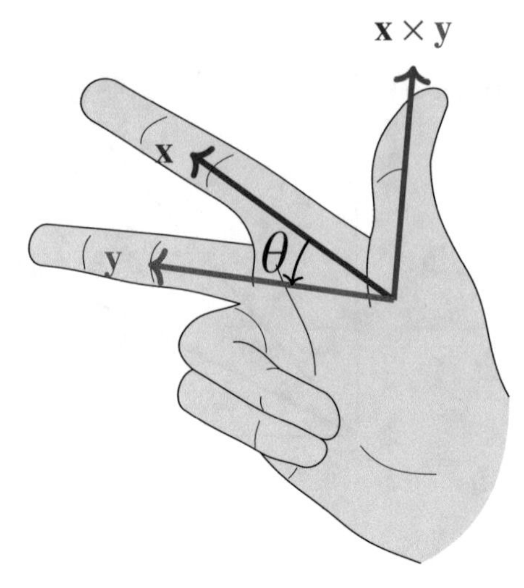

Fig. 4.5 The right-hand rule used to determined the direction of the vector $\mathbf{x} \times \mathbf{y}$.

$$\begin{cases} \mathbf{i} \times \mathbf{j} = \mathbf{k} \\ \mathbf{j} \times \mathbf{k} = \mathbf{i} \\ \mathbf{k} \times \mathbf{i} = \mathbf{j} \end{cases} \tag{4.20}$$

Note that by definition

$$\begin{cases} \mathbf{x} = x_1\mathbf{i} + x_2\mathbf{j} + x_3\mathbf{k} \\ \mathbf{y} = y_1\mathbf{i} + y_2\mathbf{j} + y_3\mathbf{k} \end{cases} \tag{4.21}$$

therefore we can calculate

$$\begin{aligned} \mathbf{x} \times \mathbf{y} =&(x_1\mathbf{i} + x_2\mathbf{j} + x_3\mathbf{k}) \times (y_1\mathbf{i} + y_2\mathbf{j} + y_3\mathbf{k}) = \\ &x_1y_1(\mathbf{i} \times \mathbf{i}) + x_1y_2(\mathbf{i} \times \mathbf{j}) + x_1y_3(\mathbf{i} \times \mathbf{k}) + \\ &x_2y_1(\mathbf{j} \times \mathbf{i}) + x_2y_2(\mathbf{j} \times \mathbf{j}) + x_2y_3(\mathbf{j} \times \mathbf{k}) + \\ &x_3y_1(\mathbf{k} \times \mathbf{i}) + x_3y_2(\mathbf{k} \times \mathbf{j}) + x_3y_3(\mathbf{k} \times \mathbf{k}) = \\ &(x_2y_3 - x_3y_2)\mathbf{i} + (x_3y_1 - x_1y_3)\mathbf{j} + (x_1y_2 - y_2x_1)\mathbf{k} \end{aligned} \tag{4.22}$$

where we have used Equations (4.20) and the fact that the cross product between parallel vectors is zero (e.g. $\mathbf{i} \times \mathbf{i} = \mathbf{0}$).

Info Division of Vectors

Note that it is not possible to define the operation of division between vectors in a unique way. First of all, the division of two vectors should be a vector (one says that the vector space should be closed with respect to the division operation). When we say "division," we really mean the inverse operation of multiplication, so that $\mathbf{x}/\mathbf{y} = \mathbf{c}$ simply means that $\mathbf{c}$ is the **unique** vector with the property $\mathbf{yc} = \mathbf{x}$ (for some definition of the multiplication operation). With this idea, we could think that we could use the just defined cross-product to also define the division. But consider the following two examples

$$\begin{aligned}(1,0,0) \times (0,1,0) = (1,0,0)\\ (1,0,0) \times (1,1,0) = (1,0,0)\end{aligned} \tag{4.23}$$

As you can see the two vectors $(0,1,0)$ and $(1,1,0)$ have both the property of giving the same result when multiplied by $(1,0,0)$. Therefore, there is no unique vector $\mathbf{y}$ with the property that $(1,0,0) \times \mathbf{y} = (1,0,0)$ and thus we cannot use the cross product to define the division.

Warning Division of Vectors in `numpy`

When developing machine learning models in Python, you will need to use the `numpy` library, as it is the *de facto* standard in numerical programming. If you have two numpy arrays `a` and `b` (of the same size) you can divide them by writing `a/b` or `numpy.divide(a,b)`.
`numpy` will return an array of the same size as `a` and `b` with components that are the ratios of the single components of `a` and `b`. Indicating the components of `a` and `b` with a_i and b_i, respectively, the resulting vector from the numpy division will have components equal to a_i/b_i. This is helpful for normalising arrays or matrices and is often used in deep learning for various tasks. Just keep in mind that this is just a practical workaround to do things efficiently, but has nothing to do with a hypothetical division operation between vectors.

4.3 Matrices

Matrices are, in their simplest form, two-dimensional arrays of numbers. In other words, a matrix is a set of numbers arranged in rows and columns that form a rectangular array. An example of 2×3 matrix is

$$\begin{pmatrix} 2 & 3 & 5 \\ 3 & 8 & 11 \end{pmatrix} \tag{4.24}$$

A generic $m \times n$ matrix A can be written as

$$A = \begin{pmatrix} a_{1,1} & a_{1,2} & \cdots & a_{1,n} \\ a_{2,1} & a_{2,2} & \cdots & a_{2,n} \\ \vdots & \vdots & \ddots & \vdots \\ a_{m,1} & a_{m,2} & \cdots & a_{m,n} \end{pmatrix} \tag{4.25}$$

where $a_{i,j}$ is the generic element at row i and columns j. In this book, we indicate matrices with uppercase letters. Here, and in machine learning in general, only matrices with real numbers as elements will be considered.

Info **Tensors**

In machine learning, you will often deal with arrays that have more than two dimensions. For example, a dataset composed of M coloured images will have 4 dimensions: two will indicate the horizontal and vertical position of a pixel, one is associated with the colour channel (if you have RGB images for example you will need 3 numbers to identify the colour of each pixel) and one with the index of the image. In general terms, an array of numbers that have more than 2 dimensions is called a **Tensor** and its elements are indicated, similarly to matrices, with subscripts. For example, a Tensor T with three dimensions will have elements indicated with $T_{i,j,k}$ where $i, j, k \in \mathbb{N}$.

One property of matrices that is important is their *shape*, or, in other words, how many elements the matrix has along each of its dimensions. For example, a matrix with 10 rows and 25 columns will have a shape given by $(10, 25)$. Typically, the shape of an array is indicated as a tuple of numbers. A generic matrix with m rows and n columns will have the shape (m, n).

4.3.1 Sum, Subtraction and Transpose

The most basic operation that you can perform on a matrix is the **transpose**. The operation simply mirrors the matrix with respect to the diagonal. In more mathematical terms the transpose of A is indicated with $A^\top$ and has elements given by

$$(A^\top)_{i,j} = A_{j,i} \tag{4.26}$$

Matrices can be added and subtracted, but only if they have the same shape. The operation happens element-wise. So for example if we have two generic matrices X and Y, then the elements of $B = X + Y$ and $C = X - Y$ will be given by

$$\begin{cases} B_{i,j} & = X_{i,j} + Y_{i,j} \\ C_{i,j} & = X_{i,j} - Y_{i,j} \end{cases} \tag{4.27}$$

Thanks to Equation (4.27) is easy to see that sum and subtraction of matrices satisfy the commutative and associative properties[4].

4.3.2 Multiplication of Matrices and Vectors

Probably the most important operation that you can do with matrices (and the one most used in machine learning) is multiplication. Note that the multiplication is only possible if the shapes of the two matrices you want to multiply satisfy a specific relationship. Suppose that we consider two matrices X with a shape of (n, m) and Y with a shape (p, q). Multiplication $C = XY$ is defined by

$$C_{i,j} = \sum_{k=1}^{m} X_{i,k} Y_{k,j} \tag{4.28}$$

and is only possible if $m = p$. For example, to calculate the element $C_{2,2}$ you will need to use the second row of the matrix X (you will need all elements $X_{2,j}$) and the second column of the matrix Y (all elements $Y_{j,2}$). In Figure 4.6 you can see a graphical representation of the multiplication process between the two matrices X (with shape (n, m)) and Y (with shape (m, q)). To obtain the element $C_{2,2}$ one has to multiply the second row elements of X (marked blue in Figure 4.6) **element-wise** with the second column elements of Y (marked orange in Figure 4.6) as explained by the red marked elements in Figure 4.6 and then sum the results of the element-wise multiplications.

Multiplication is distributive

$$X(Y + Z) = XY + XZ \tag{4.29}$$

and associative

$$X(YZ) = (XY)Z \tag{4.30}$$

It is important to note that matrix multiplication is **not** commutative. So, in general,

$$XY \neq YZ \tag{4.31}$$

The transpose of the product of two matrices XY can be calculated with the formula

$$(XY)^\top = Y^\top X^\top \tag{4.32}$$

Matrices can be multiplied with vectors, of course. We can use Equation (4.28) by writing a vector $\mathbf{y} = (y_1, ..., y_m)$ as a matrix with 1 column and m row (shape $(m, 1)$) (or intuitively in vertical notation)

[4] The commutative property says that changing the order of the elements you are summing does not change the result, or in other words $a + b = b + a$. The associative property says that if you are summing, for example, three elements a, b and c that $(a + b) + c = a + (b + c)$.

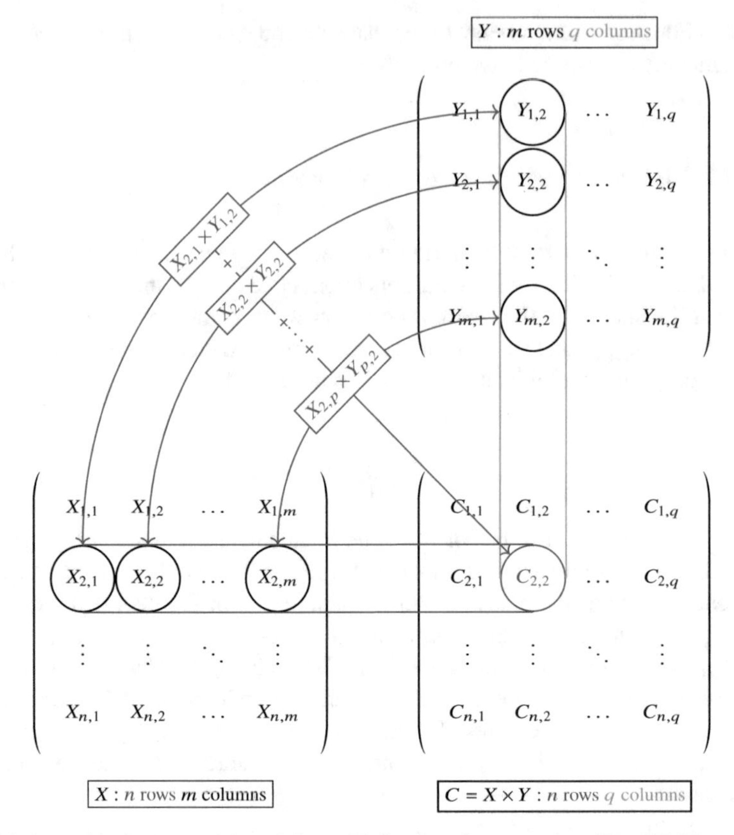

Fig. 4.6 A graphical representation of the multiplication of two matrices X and Y. The matrix C is given by $C = XY$. To obtain the element $C_{2,2}$ one has to multiply the 2nd row elements from X (marked in blue) **element-wise** with the second column elements of Y (marked in orange) as explained by the red marked elements and then sum the results of the element-wise multiplications.

$$\mathbf{y} = \begin{pmatrix} y_1 \\ \vdots \\ y_m \end{pmatrix} \tag{4.33}$$

by doing that, if we have a matrix A with a shape of (n, m) we can easily calculate the product by using Equation (4.28) obtaining

$$A\mathbf{y} = \begin{pmatrix} a_{1,1} & a_{1,2} & \cdots & a_{1,m} \\ a_{2,1} & a_{2,2} & \cdots & a_{2,m} \\ \vdots & \vdots & \ddots & \vdots \\ a_{n,1} & a_{n,2} & \cdots & a_{n,m} \end{pmatrix} \begin{pmatrix} y_1 \\ y_2 \\ \vdots \\ y_m \end{pmatrix} = \begin{pmatrix} a_{1,1}y_1 + \cdots a_{1,m}y_m \\ a_{2,1}y_1 + \cdots a_{2,m}y_m \\ \cdots \\ a_{n,1}y_1 + \cdots a_{n,m}y_m \end{pmatrix} \tag{4.34}$$

4.3.3 Inverse and Trace

The inverse of a matrix X is indicated with X^{-1} and is the matrix that satisfy the equation

$$X^{-1}X = I \tag{4.35}$$

where I is the identify matrix of shape (n, n) as the matrix that satisfy the condition

$$\forall \mathbf{x} \in \mathbb{R}^n \; I\mathbf{x} = \mathbf{x} \tag{4.36}$$

This is a convoluted way of defining a matrix that has all the diagonal elements equal to 1 and all the other equal to 0. For example a $(3, 3)$ identify matrix looks like this

$$\begin{pmatrix} 1 & 0 & 0 \\ 0 & 1 & 0 \\ 0 & 0 & 1 \end{pmatrix} \tag{4.37}$$

Note that a matrix X has an inverse X^{-1} only if it is **square** and all columns are **linearly independent**. We will not spend time here in discussing those conditions, but it is good to know that not all matrices can be inverted.

Info System of Linear Equations and Inverse of Matrices

In general a linear set of equations

$$\begin{cases} a_{1,1}x_1 + \cdots a_{1,m}x_m = b_1 \\ a_{2,1}x_1 + \cdots a_{2,m}x_m = b_2 \\ \ldots \\ a_{m,1}x_1 + \cdots a_{m,m}x_m = b_m \end{cases} \tag{4.38}$$

can be written in what is called **matrix form** as

$$A\mathbf{x} = \mathbf{b} \tag{4.39}$$

where

$$\begin{cases} A_{i,j} & = a_{i,j} \\ (\mathbf{x})_i & = x_i \\ (\mathbf{b})_i & = b_i \end{cases} \tag{4.40}$$

By looking at Equation (4.39) it is to see that a solution $\mathbf{x}$ is simply given by

$$\mathbf{x} = A^{-1}\mathbf{b} \tag{4.41}$$

under the assumption that A^{-1} exists. Solving a system of linear equations with this approach is much faster than by using the substitution method (deriving an expression for x_1 from one of the equations and substituting it into the others, then deriving x_2 and so on). If the matrix A^{-1} does not exist, the system has no solution and vice versa (the discussion of the existence of the inverse goes beyond the scope of this book). For example, the system

$$\begin{cases} x_1 - x_2 = 0 \\ -x1 + x2 = 3 \end{cases} \tag{4.42}$$

has clearly no solutions (you should verify it). Analogously the matrix

$$A = \begin{pmatrix} 1 & -1 \\ -1 & 1 \end{pmatrix} \tag{4.43}$$

is not invertible. In the next section, we will see how to realise if a matrix is invertible.

The trace of a matrix X is defined only for square matrices and is indicated by $\mathrm{Tr}(X)$. It is simply the sum of the elements on the diagonal

$$\mathrm{Tr}(X) = \sum_{i=1}^{n} X_{i,i} \tag{4.44}$$

The trace of matrix is rarely used in applied machine learning nonetheless is good to know its main properties.

$$\begin{cases} \mathrm{Tr}(A + B) = \mathrm{Tr}(A) + \mathrm{Tr}(B) \\ \mathrm{Tr}(cA) = c\mathrm{Tr}(A); \quad c \in \mathbb{R} \\ \mathrm{Tr}(A^\top B) = \mathrm{Tr}(AB^\top) = \mathrm{Tr}(B^\top A) = \mathrm{Tr}(BA^\top) \\ \mathrm{Tr}(AB) = \mathrm{Tr}(BA) \\ \mathrm{Tr}(ABCD) = \mathrm{Tr}(BCDA) = \mathrm{Tr}(CDAB) = \mathrm{Tr}(DABC) \ \text{(cyclic property)} \end{cases} \tag{4.45}$$

4.3.4 ★ Determinant

For a square matrix X we can define the **determinant**: a function that maps matrices to scalars. It is indicated with $\det(X)$ or $|X|$. Before looking at how to calculate it and providing a formula, it is highly educational to grasp its meaning. Here I will provide a geometrical interpretation of it. Let us suppose we have a matrix

$$\begin{pmatrix} \text{---} \ \mathbf{x}_1^\top \ \text{---} \\ \vdots \\ \text{---} \ \mathbf{x}_n^\top \ \text{---} \end{pmatrix} \tag{4.46}$$

where $\mathbf{x}_i^\top$ indicates the ith matrix row. Now consider the set S of all points formed by taking all possible linear combinations of the vectors $\mathbf{x}_i^\top$ for $i = 1, ..., n$ with coefficients $\alpha_i \in]0, 1]$. This can be expressed in more mathematical form

$$S = \{p \in \mathbb{R}^n | p = \sum_{i=1}^{n} \alpha_i \mathbf{x}_i^\top \text{ where } \alpha_i \in]0, 1], i = 1, ..., n\} \tag{4.47}$$

The absolute value of $\det(X)$ is the measure of the volume of the set S. We have not formally defined what volume means in this context, as this would go beyond the scope of this book, but an example in two dimensions will give the student an adequate intuitive understanding of this concept. Consider a generic matrix $X \in \mathbb{R}^{2\times 2}$.

$$X = \begin{pmatrix} X_{1,1} & X_{1,2} \\ X_{2,1} & X_{2,2} \end{pmatrix} \tag{4.48}$$

in this case we will have

$$\begin{cases} \mathbf{x}_1^\top & = (X_{1,1} \ \ X_{1,2}) \\ \mathbf{x}_2^\top & = (X_{2,1} \ \ X_{2,2}) \end{cases} \tag{4.49}$$

In Figure (4.7) you can see two vectors in a Cartesian space marked in blue ($\mathbf{x}_1^\top$) and in red ($\mathbf{x}_2^\top$). The figure will help us in understanding visually the calculations below. Note that all the formulas we will write are completely general and do not depend on the particular vectors in Figure 4.7.

The set S that we have defined in Equation (4.47) is, for the vectors in Figure 4.7, the shaded region in Figure (4.7). Now, let us calculate the area A of the shaded area in Figure (4.7) for our generic matrix X. In general A is given by the formula[5]

$$A = |\mathbf{x}_1^\top|_2 h = |\mathbf{x}_1^\top|_2 |\mathbf{x}_2^\top|_2 \sin\theta \tag{4.50}$$

where h is the segment indicated with a dashed line in Figure 4.7. We now only need to derive a formula for the angle θ. We can do this by using Equation (4.16)

$$\cos\theta = \frac{\mathbf{x}_1^\top \cdot \mathbf{x}_2^\top}{|\mathbf{x}_1^\top|_2 |\mathbf{x}_2^\top|_2} \tag{4.51}$$

Now we have all ingredients to derive a formula for A. In fact,

[5] You should remember how to calculate the area of a parallelogram.

$$
\begin{aligned}
A &= |\mathbf{x}_1^\top|_2 |\mathbf{x}_2^\top|_2 \sin\theta = |\mathbf{x}_1^\top|_2 |\mathbf{x}_2^\top|_2 \sqrt{1-\cos\theta^2} = \\
&= |\mathbf{x}_1^\top|_2 |\mathbf{x}_2^\top|_2 \frac{\sqrt{|\mathbf{x}_1^\top|_2^2 |\mathbf{x}_2^\top|_2^2 - (\mathbf{x}_1^\top \cdot \mathbf{x}_2^\top)^2}}{|\mathbf{x}_1^\top|_2 |\mathbf{x}_2^\top|_2} = \\
&= \sqrt{(X_{1,1}^2 + X_{1,2}^2)(X_{2,1}^2 + X_{2,2}^2) - (X_{1,1}X_{2,1} + X_{1,2}X_{2,2})^2} = \\
&= \{\text{With some work}\} = \\
&= |X_{1,1}X_{2,2} - X_{1,2}X_{2,1}|
\end{aligned}
\tag{4.52}
$$

Note that technically speaking the square root of a square can be positive or negative, but since we are looking at the area the result must be positive. As you will see in

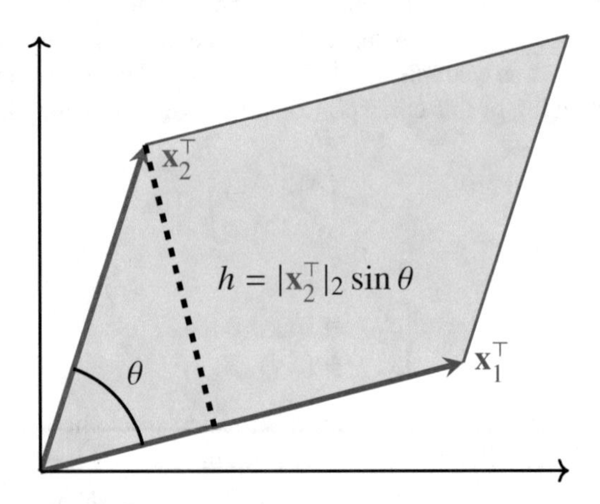

Fig. 4.7 The vectors $\mathbf{x}_1^\top$ and $\mathbf{x}_2^\top$ in a cartesian space marked in blue and in red respectively. The shaded area contains all the elements of the set S that we have defined in Equation (4.47). h indicates the length of the component of the vector $\mathbf{x}_2^\top$ along the direction perpendicular to $\mathbf{x}_1^\top$.

what follows, the last line in Equation (4.52) is the absolute value of the determinant of X.

Let us now give a formal definition of the determinant for a 2×2 and a 3×3 matrix.

Definition 4.2 The determinant of a generic matrix $X \in \mathbb{R}^{2\times 2}$ is given by

$$
\det(X) = \det \begin{pmatrix} X_{1,1} & X_{1,2} \\ X_{2,1} & X_{2,2} \end{pmatrix} = X_{1,1}X_{2,2} - X_{1,2}X_{2,1} \tag{4.53}
$$

Definition 4.3 For a 3×3 matrix the determinant is given by this formula

$$\det(X) = \det\begin{pmatrix} X_{1,1} & X_{1,2} & X_{1,3} \\ X_{2,1} & X_{2,2} & X_{2,3} \\ X_{3,1} & X_{3,2} & X_{3,3} \end{pmatrix} = X_{1,1}\det\begin{pmatrix} X_{2,2} & X_{2,3} \\ X_{3,2} & X_{3,3} \end{pmatrix} + \\ +X_{1,2}\det\begin{pmatrix} X_{2,1} & X_{2,3} \\ X_{3,1} & X_{3,3} \end{pmatrix} + X_{1,3}\det\begin{pmatrix} X_{2,1} & X_{2,2} \\ X_{3,1} & X_{3,2} \end{pmatrix} \tag{4.54}$$

the final result can then be calculated by evaluating the determinants of the 2×2 matrices. Each of those determinants is called a *minor* of the matrix X.

The determinants for matrices that larger than 3×3 is defined analogously in a recursive fashion. This is called the Laplace expansion, and the general recursive formula for the determinants of an arbitrary matrix X of shape (n, n) is given by

$$\det(X) = \sum_{j=1}^{n} (-1)^{i+j} X_{i,j} M_{i,j} \tag{4.55}$$

where $M_{i,j}$ is the determinant of the submatrix obtained by removing the ith row and the jth column of X. In the formula one has to choose one value for i, for example $i = 1$. Note that we report this formula just for completeness and for large matrices numerical approaches are almost always used in practice, but this expansion is often used in mathematical proofs and formulas.

Info ★ The Role of the Determinant in the Inversion of a Matrix

Earlier we briefly discussed that not all matrices can be inverted. In general, the inverse of a matrix X^{-1} is proportional to the inverse of the determinant[a]

$$X^{-1} \propto \frac{1}{\det(X)} \tag{4.56}$$

and therefore the inverse of a matrix exists if and only if the determinant is not zero. This is an easy way of verifying if one matrix can be inverted. In the example we have looked at before when we tried to solve the system of linear equations (4.42) we had to invert the matrix

$$A = \begin{pmatrix} 1 & -1 \\ -1 & 1 \end{pmatrix} \tag{4.57}$$

You may remember that we said that this matrix is not invertible. We can verify this statement easily by calculating the determinant

$$\det(A) = a_{1,1}a_{2,2} - a_{1,2}a_{2,1} = (1)(1) - (-1)(-1) = 0 \tag{4.58}$$

and as expected, since the determinant is equal to zero, the matrix A^{-1} does not exist.

[a] We will not prove this here.

4.3.5 ★ Matrix Calculus and Linear Regression

Matrix calculus is the study of calculus with matrices. Is not obvious how to perform a derivative, for example, of a matrix or even less obvious how to perform it with respect to a matrix (or a vector, for that matter). This short section intends to give the reader an overview of the most important definitions and show with an example (linear regression) how powerful these techniques are when used correctly. Let us start with a few definitions.

Definition 4.4 If we consider a matrix X of shape (n, m) that depends on some scalar variable y, we define its derivative with respect to y with

$$\frac{dX}{dy} = \begin{pmatrix} dX_{1,1}/dy & \cdots & dX_{n,1}/dy \\ \vdots & \ddots & \vdots \\ dX_{n,1}/dy & \cdots & dX_{n,m}/dy \end{pmatrix} \tag{4.59}$$

Analogously we can define the derivative of a vector $\mathbf{x}$ that depends on some scalar y.

Definition 4.5 The derivative of a vector $\mathbf{x}$ of shape $(n, 1)$ with respect to some scalar y is defined by

$$\frac{d\mathbf{x}}{dy} = \begin{pmatrix} dx_1/dy \\ \vdots \\ dx_n/dy \end{pmatrix} \tag{4.60}$$

Now, let us turn our attention to the derivative of a function with respect to a matrix or a vector. Consider a function $f(X)$ and that X is a matrix of shape (n, m). The following definitions are common.

Definition 4.6 The derivative of a scalar function with respect to a matrix X and a vector $\mathbf{x}$ are given by

$$\frac{df}{dX} = \begin{pmatrix} df/dX_{1,1} & \cdots & df/dX_{n,1} \\ \vdots & \ddots & \vdots \\ df/dX_{n,1} & \cdots & df/dX_{n,m} \end{pmatrix} \tag{4.61}$$

and

$$\frac{df}{d\mathbf{x}} = (df/dx_1, \cdots, df/dx_n) \tag{4.62}$$

respectively. Note that $df/d\mathbf{x}$ is nothing more than the gradient of a function that we have already seen (we have indicated with $\nabla_{\mathbf{x}} f$ so far). The gradient of a function is one of the key and important elements when training neural networks and was discussed in length in 3.5.

All this new notation may seem an unnecessary complication, but it is extremely powerful and can be used in many practical cases. To give the student an idea of its power, let us discuss the problem of linear regression with least squares. Let us consider M tuples of real numbers $(x^{(i)}, y^{(i)})$ for $i = 1, ..., M$, $x^{(i)} \in \mathbb{R}^n$ and $y^{(i)} \in \mathbb{R}$. Our goal is to find the best parameters θ_i for $i = 0, ..., n$ that minimise the mean square error (MSE) between the formula

$$\hat{y}^{(j)} = \sum_{i=1}^{n} \theta_i x_i^{(j)} + \theta_0 \tag{4.63}$$

and the expected value $y^{(i)}$. The MSE in this case can be written as

$$J = \frac{1}{n}\sum_{i=1}^{n}(y^{(i)} - \hat{y}^{(i)})^2 = \frac{1}{n}\sum_{i=1}^{n}\left(y^{(i)} - \sum_{j=1}^{n}\theta_i x_i^{(j)} - \theta_0\right)^2 \tag{4.64}$$

To solve this problem we need to minimise J with respect to all θ_i. This can be done directly by using Equation (4.64) and calculating its derivatives with respect to θ_i of course, but their form will be quite complicated to handle. But luckily matrices are here to save us. Let us first rewrite Equation (4.63) in matrix form by defining first the vector $\boldsymbol{\theta} = (\theta_0, \theta_1, ..., \theta_n)$ (with shape $(n+1, 1)$). Since we have a constant term in Equation (4.63), we define the matrix

$$X = \begin{pmatrix} 1 & x_1^{(1)} & \cdots & x_n^{(1)} \\ 1 & x_1^{(2)} & \cdots & x_n^{(2)} \\ \vdots & \vdots & \ddots & \vdots \\ 1 & x_1^{(m)} & \cdots & x_n^{(m)} \end{pmatrix} \tag{4.65}$$

by adding a column with all 1s to the matrix obtained by putting all $x^{(i)}$ together. Note that X have a shape equal to $(m, n+1)$. With those definitions we can finally rewrite Equation (4.63) in matrix form

$$\hat{\mathbf{y}} = X\boldsymbol{\theta} \tag{4.66}$$

where $\hat{\mathbf{y}} = (\hat{y}_1, ..., \hat{y}_n)^\top$. Now let us rewrite J from Equation (4.64) with matrices

$$J = \frac{1}{m}(X\boldsymbol{\theta} - \mathbf{y})^\top (X\boldsymbol{\theta} - \mathbf{y}) \tag{4.67}$$

This equation is much more easy to handle than Equation (4.64). Let's expand it

$$\begin{aligned} J =& \frac{1}{m}(\mathbf{y} - X\boldsymbol{\theta})^\top(\mathbf{y} - X\boldsymbol{\theta}) = \frac{1}{m}(\mathbf{y}^\top - \boldsymbol{\theta}^\top X^\top)(\mathbf{y} - X\boldsymbol{\theta}) = \\ =& \frac{1}{m}\left(\boldsymbol{\theta}^\top X^\top X\boldsymbol{\theta} - \boldsymbol{\theta}^\top X^\top \mathbf{y} - \mathbf{y}^\top X\boldsymbol{\theta} + \mathbf{y}^\top\mathbf{y}\right) \end{aligned} \tag{4.68}$$

all the elements that do not contain the matrix $\boldsymbol{\theta}$ are not relevant when we will calculate $\partial J/\partial\boldsymbol{\theta}$ and therefore will be ignored in the next calculations. Now before proceeding I will cheat and give you some formulas without derivation (all the formulas can be found in [7]).

$$\frac{\partial\boldsymbol{\theta}^\top A\boldsymbol{\theta}}{\partial\boldsymbol{\theta}} = (A + A^\top)\boldsymbol{\theta} \tag{4.69}$$

$$\frac{\partial\boldsymbol{\theta}^\top \mathbf{a}}{\partial\boldsymbol{\theta}} = \mathbf{a} \tag{4.70}$$

$$\frac{\partial\mathbf{a}^\top \boldsymbol{\theta}}{\partial\boldsymbol{\theta}} = \mathbf{a} \tag{4.71}$$

by using Equations (4.69), (4.70) and (4.71) in Equation (4.68) it is possible to derive the formula

$$\begin{aligned} \frac{\partial J}{\partial\boldsymbol{\theta}} =& \frac{1}{m}\frac{\partial}{\partial\boldsymbol{\theta}}\left(\underbrace{\boldsymbol{\theta}^\top X^\top X\boldsymbol{\theta}}_{\text{use 4.69}} - \underbrace{\boldsymbol{\theta}^\top X^\top \mathbf{y}}_{\text{use 4.70}} - \underbrace{\mathbf{y}^\top X\boldsymbol{\theta}}_{\text{use 4.71}} + \mathbf{y}^\top\mathbf{y}\right) = \\ =& \frac{2}{m}(X^\top X\boldsymbol{\theta} - X^\top\mathbf{y}) \end{aligned} \tag{4.72}$$

Now we want to minimise J therefore we have to impose

$$\frac{\partial J}{\partial\boldsymbol{\theta}} = 0 \tag{4.73}$$

with that and Equation (4.72) we finally get the result

$$\boldsymbol{\theta} = (XX^\top)^{-1}X^\top\mathbf{y} \tag{4.74}$$

Equation (4.74) gives the values of $\boldsymbol{\theta}$ that minimise the MSE.

Let us now use this formula and compare it to the numerical solution obtained by minimising the MSE. Let us generate 30 tuples $(x^{(i)}, y^{(i)})$ with $i = 1, ..., 30$ with the equation

$$y^{(i)} = 2x^{(i)} + 1 + \epsilon^{(i)} \tag{4.75}$$

by choosing $x^{(i)}$ randomly between 0 and 10 (from a uniform distribution) and where $\epsilon^{(i)}$ is chosen randomly by sampling from a normal distribution with average equal to zero and variance equal to 1. The data is shown in Figure 4.8. The red points have

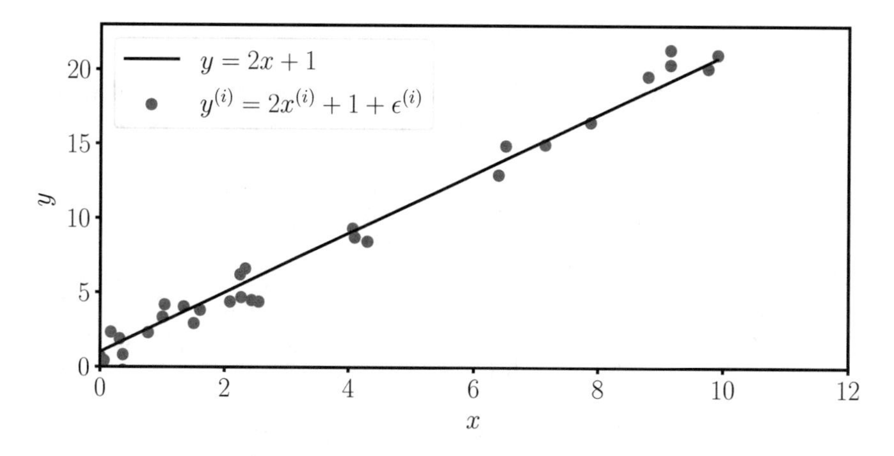

Fig. 4.8 Sample data obtained by adding noise (red points) to a linear function $2x + 1$.

been obtained as explained, while the black lines has been obtained with Equation (4.75) without $\epsilon^{(i)}$. Note that if you decide to recreate this dataset, your points $y^{(i)}$ will be different due to a different random seed in your numerical routines. So don't be surprised if this happens. If we use normal numerical routines[6] to minimise the MSE we get, for the data in Figure 4.8, the value for the slope of 2.096 and for the intercept of 0.520. Now if we use our exact formula we get the results

$$\boldsymbol{\theta} = (XX^\top)^{-1}X^\top \mathbf{y} = \begin{pmatrix} 0.520 \\ 2.096 \end{pmatrix} \tag{4.76}$$

that are exactly the same results we obtained with the numerical approach. By using matrix formalism, we can simplify the notation extremely and gets formula that are much easier to handle than summation over large number of elements.

4.4 Relevance for Machine Learning

Matrices are used extensively in machine learning and especially in deep learning. The weights in FFNNs or the filters in CNNs are always matrices (typically very large) and often they are tensors since they have more than two dimensions. Matrices are one of the fundamental building blocks of deep learning. The input data are also always saved as a tensor. Imagine that you have M images, each having a resolution of 1024×1024 in RGB (3 channels). Then the entire dataset can be saved as a huge tensor that has 4 dimensions: index of the image (M values), pixel index along the x axis (1024 values), pixel index along the y axis (1024 values), 3 values for the colour

[6] If you are interested, to obtain these results I have used Python and scikit-learn.

in RGB (Red value, Green value and Blue value, so 3 values). The dataset tensor has four dimensions and shape (M, 1024, 1024, 3).

This is why the Python `numpy` library is so commonly used. It was developed exactly with the goal of performing operations on arrays of numbers in the most efficient way possible.

Info GPUs, TPUs and Matrices

GPUs (Graphical Processing Units) are very good at multiplying matrices. In three-dimensional graphics, matrices are fundamental, since they are used to (for example) rotate objects in space. Graphic cards were built to be very fast at that: multiplying matrices. This is the reason why they are so useful in deep learning. At its core, training a deep learning model consists of numerous matrix multiplication operations. The faster you can multiply tensors, the faster you can train a model. At the time of writing, the only GPUs that can be used for deep learning are NVIDIA ones (with possibly the exception of GPUs in M1/M2/M3 apple chipsets). Since 2007 the NVIDIA CUDA framework provides access to GPU resources. CUDA is based on C and provides an API that can be used to access GPUs for machine learning tasks. Since 2021, Apple with their M1 chip laptops managed to make their GPUs available for deep learning with the Metal framework (for example with TensorFlow), but the NVIDIA cards are found in all servers and deep learning workstations. Note also that CUDA works on Linux and windows and is not available for MacOs. An interesting review of methods to optimise code for GPUs can be found in [8].

An additional advantage of using GPUs is the fact that they can scale when used in parallel (which means that you can use multiple GPUs to speed up your training) and that comes very often with a large amount of memory for very large datasets. 48 Gb or more of memory for each graphic card are not uncommon in the more high end segment. Note that such cards can cost up to (and more than) 10'000 USD. Setting up a professional deep learning infrastructure can be very expensive.

If your models are large, but you do not want to invest in an expensive deep learning infrastructure, you can always use cloud solutions such as Google Cloud, Microsoft Azure or AWS (Amazon Web Services). They provide a pay-by-use infrastructure that can be dynamically changed according to your needs.

TPUs (Tensor Processing Units) are special chips developed by Google that can do one thing, and one thing only: multiply matrices. They cannot run a word processor or a text editor, but they can multiply matrices at a very fast speed [9, 10]. They are available on Google Cloud and they provide quite a speed up to model training. Those are not available to buy for workstations but are used in Google Linux Clusters.

4.5 Eigenvectors and Eigenvalues

Eigenvectors and eigenvalues are fundamental concepts in linear algebra associated with square matrices. Given a square matrix A, an eigenvector $\mathbf{v}$ is a non-zero vector that, when multiplied by A, results in a vector that is a scalar multiple of $\mathbf{v}$. This relationship is expressed by the equation

$$A\mathbf{v} = \lambda\mathbf{v} \tag{4.77}$$

where λ is a scalar known as the eigenvalue corresponding to the eigenvector $\mathbf{v}$. In essence, the action of the matrix A on its eigenvector $\mathbf{v}$ is to simply stretch or compress it by the factor λ. This concept is crucial in many areas of mathematics and engineering, including systems theory, quantum mechanics, and principal component analysis in statistics (more on this in Section 4.6). The theory revolving around eigenvectors and eigenvalues is rich and complex, and it goes well beyond this book, but it is useful at least to have an intuitive understanding of what they are and why they are relevant. To do that, let us introduce some useful terminology.

Definition 4.7 Consider A and B two $n \times n$ matrices. We will call A and B **similar** if there exists an invertible matrix P such that

$$A = PBP^{-1} \tag{4.78}$$

Note that if A is similar to B, then B is similar to A. In fact, if 4.78 is true, then it is easy to see that

$$B = P^{-1}AP \tag{4.79}$$

If A and B are similar one can prove (we will not do that here) that the determinant of the two matrices is the same, and the A and B have the same eigenvalues. Now we are getting closer to the interesting part.

Definition 4.8 A matrix A is called *diagonalisable* if it is similar to a diagonal matrix D. In other words, if a matrix P exists such that

$$A = PDP^{-1} \tag{4.80}$$

It can be proven that, under certain assumptions, if a matrix is diagonalisable then D will have as diagonal elements the eigenvalues of A and zero for all other elements. So the challenge at this point is, given a matrix A, how to find P and D. It turns out that, again under certain assumptions, P is a matrix composed of the eigenvectors of A, and D is a diagonal matrix with all the eigenvalues of A. Therefore, the challenge is to find the eigenvectors and eigenvalues of a generic matrix A. This problem is relevant since, for example, it is one way to solve the problem formulated as principal component analysis (PCA), as described in Section 4.6.

Discussing all the theory behind eigenvectors and eigenvalues would go beyond the scope of this book, and any good linear algebra book will contain all the information you will need. For the purposes of machine learning, an intuitive understanding of what they are will suffice. In any practical problem, you will not need to determine neither eigenvectors nor eigenvalues.

Warning ★ Not all Matrices are Diagonalizable

It is important to remark that not every square matrix is diagonalisable. A matrix is said to be *non-diagonalizable* or *defective* when it does not have a complete set of linearly independent eigenvectors. This situation arises when the algebraic multiplicity of an eigenvalue (the number of times it appears as a root of the characteristic equation) is greater than the geometric multiplicity (the dimension of the eigenspace corresponding to that eigenvalue). Consider the matrix:

$$A = \begin{pmatrix} 4 & 1 \\ 0 & 4 \end{pmatrix}$$

Both eigenvalues of this matrix are 4. Furthermore, the eigenvectors corresponding to this eigenvalue form a one-dimensional space, making it impossible to construct a basis for $\mathbb{R}^2$ using these eigenvectors alone. Therefore, matrix A is non-diagonalisable. Another example is the matrix:

$$B = \begin{pmatrix} 2 & 1 \\ 0 & 2 \end{pmatrix}$$

Like the previous example, both eigenvalues are 2, but there is only one linearly independent eigenvector associated with this eigenvalue. Thus, matrix B is also non-diagonalisable. In summary, matrices that lack a sufficient number of linearly independent eigenvectors to span the vector space they operate in are non-diagonalisable.

With this intuitive understanding of what eigenvectors and eigenvalues are, we can turn our attention to PCA. This is a wonderful demonstration of how powerful linear algebra is, and how much machine learning depends on it. Note that the mathematics necessary to understand PCA is slightly more complex than our discussion so far.

4.6 ★ Principal Component Analysis

Principal Component Analysis (PCA) is a widely used method to extract relevant and sometimes hidden information in complex datasets. Very often, depending on how data collection is done (we will give an example later), we end up with many more variables and information than what is really relevant to the problem. PCA is called a *dimensionality reduction* technique because by extracting only relevant information, it effectively reduces the number of dimensions of a problem ignoring redundant and irrelevant information.

To better understand what is meant, suppose that we want to study the movement of an object attached to a spring[7]. Assuming the right initial conditions, we know from physics that the movement will be approximately (the spring may not be ideal, the surface on which the experiment is done not completely flat, the initial condition not perfect, etc.) along a line. Let us suppose that we do not know much about the problem and we decide to study it with three cameras positioned randomly around the system (spring plus object). Let us also assume that each camera measures the position of the object at regular intervals. We already know that three cameras are too much, and one would be more than enough. However, by redundantly setting up our measurement system, we generate data with a much higher dimensionality than necessary. Very often, as experimenters, we often have no idea about which measurement setup reflect in the best possible way how a specific system works.

The question we are trying to address is *can we remove the redundant information and extract the physical relevant data from our measurements*? An additional problem that makes our life more difficult is that in real life, systems and measurement systems are imperfect. This means that in our example, we will not measure a perfect line of points. The spring may be imperfect, the object not completely symmetric, friction on the surface may be irregular and make the spring oscillate in multiple directions, the movement will slow down with time, and so on. This means that we will not observe a perfect line of points, but a cloud (more or less spread) depending on a variety of factors that we cannot always control.

4.6.1 Basis of a Vector Space

Let us first briefly discuss some necessary mathematical formalism. First, consider a **vector space** V. Intuitively, a vector space V is a set of vectors. For example, the entire set of all bidimensional vectors (v_1, v_2) (in other words, vectors with two components). The formal definition is more complex but is not necessary for our discussion (which I will give it in the next section, which can be safely skipped without compromising understanding). Secondly, let us define what a basis of a vector space is.

Definition 4.9 (Definition of a basis) A basis of a vector space V, is a set B of vectors such that every vector in V can be written in a unique way as a linear combination of vectors in B.

In vector notation, a **basis** is described by a set of unit vectors (in other words, each having a length of one). An example of a basis in $\mathbb{R}^2$ is given by the vectors $(1, 0)$ and $(0, 1)$. Note that the vectors of a basis do not need to be orthonormal. For example, the vectors $(1, 0)$ and $(1, 1)$ form a basis for $\mathbb{R}^2$ but they are not orthonormal. We must determine how a transformation of basis can be represented using matrix notation. Note that we will consider here only a *linear* basis transformation considering only a linear combination of the existing unit vectors that describe our original basis.

[7] Example adapted from [11].

4.6.2 ★ Definition of a Vector Space

To define a vector space, we first need to define a field.

Definition 4.10 A field $\mathbb{F}$ is a set of numbers, such that if $a, b \in \mathbb{F}$ then $a + b, a - b, a/b \in \mathbb{F}$, assuming $b \neq 0$ in the division.

Examples of fields are $\mathbb{R}, \mathbb{N}, \mathbb{Z}$, etc. An example of a set that is **not** a field is $\{1, 5\}$, as 1-5 is not in the set. Now we can define a vector space V.

Definition 4.11 A vector space V, consists in a set of elements called vectors, a field $\mathbb{F}$ of elements called scalars, and two operations.

- **Addition**: that takes two vectors $v, w \in V$ and produce a vector $v + w \in V$.
- **Scalar Multiplication**: that takes one vector $v \in V$, and a scalar $a \in \mathbb{F}$ and produces a vector $av \in V$.

The operations must satisfy the following axioms.

- **Associativity of addition**: $(u + v) + w = u + (v + w)$, with $u, v, w \in V$.
- **Zero vector**: in V exist a **zero** vector $\mathbf{0}$ such that $v + \mathbf{0} = v \; \forall v \in V$.
- **Negative vector**: in V exist, $\forall v \in V$, a vector $-v$ such that $v - v = \mathbf{0}$.
- **Associativity of multiplication**: $(ab)u = a(bu) \; \forall a, b \in \mathbb{F}, u \in V$.
- **Distributivity**: $(a + b)u = au + bu \; \forall a, b \in \mathbb{F}, u \in V$.
- **Unitarity**: $1u = u \; \forall u \in V$.

As you may notice, the definition is quite lengthy, but not necessary for a first understanding of PCA.

4.6.3 ★ Linear Transformations (maps)

In general a linear transformation (also called *map*) f from a vector space V into itself $f : V \to V$, is one that satisfies the two properties

1. **Additivity**: $f(x + y) = f(x) + f(y)$
2. **Homogeneity**: $f(cx) = cf(x)$

There is an important theorem on linear transformations that is quite relevant to our discussion.

Theorem 4.2 *(a) Every linear transformation f between finite-dimensional vector spaces can be obtained by multiplication with a unique matrix. (b) Matrix multiplications are linear transformations.*

Proof Let us first prove part (a) for a linear transformation between $V = \mathbb{R}^n$ and $V = \mathbb{R}^n$. V have, by hypothesis, a basis that we can indicate with $(\mathbf{v}_1, ..., \mathbf{v}_n)$. Now consider a linear transformation $f : V \to V$. We can write for a generic vector $\mathbf{v} \in V$

$$\mathbf{v} = \alpha_1 \mathbf{v}_1 + \cdots + \alpha_n \mathbf{v}_n \tag{4.81}$$

and that, assuming we are using the basis given by the vectors $\mathbf{v}_i$, $\mathbf{v}$ will be simply indicated by its components $\mathbf{v} = (\alpha_1, \cdots, \alpha_n)$. Now we can write for $f(v)$

$$\mathbf{u} \equiv f(\mathbf{v}) = \alpha_1 f(\mathbf{v}_1) + \cdots \alpha_n f(\mathbf{v}_n) \tag{4.82}$$

or in matrix notation

$$\mathbf{v} = (f(\mathbf{v}_1) \cdots f(\mathbf{v}_n)) \begin{pmatrix} \alpha_1 \\ \vdots \\ \alpha_n \end{pmatrix} \tag{4.83}$$

where $(f(\mathbf{v}_1), \cdots, f(\mathbf{v}_n))$ is a matrix where the ith column is given by the vector $f(\mathbf{v}_i)$. This concludes the proof of (a). The proof of (b) is trivial, as matrix multiplication is a linear transformation (and if you are not convinced you can easily verify that). □

This is highly relevant to our discussion since **PCA does nothing else than expressing the original data on a new basis that has been obtained by a linear combination of the original basis vectors.**

In general, to make a *linear* change of basis, it suffices to multiply our identity matrix (our original basis) by a **transformation matrix** T. Let us give an example, to make the discussion more concrete. Let us consider the Euclidean space $\mathbb{R}^2$ with basis vectors $v_1 = (1, 0)$ and $v_2 = (0, 1)$. Let us suppose that we want to rotate the axis of an angle α. It is easy to see, with some trigonometry, that the new vectors u_1 and u_2, obtained by rotating v_1 and v_2, are

$$u_1 = (\cos\alpha, \sin\alpha) \tag{4.84}$$

$$v_1 = (-\sin\alpha, \cos\alpha) \tag{4.85}$$

The formula can be easily understood by looking at Figure 4.9. It is then easy to

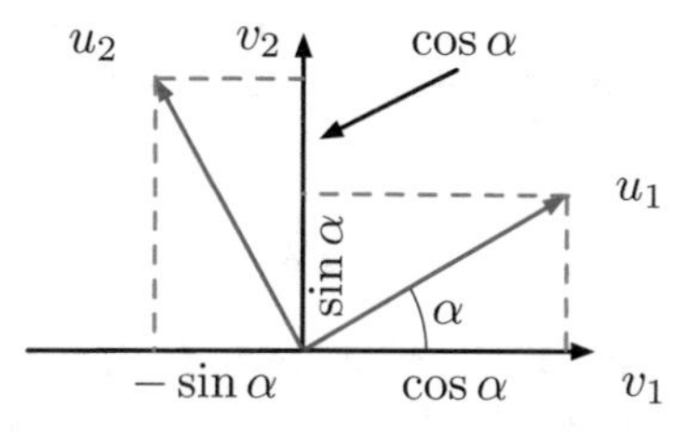

Fig. 4.9 A change of basis from (v_2, v_2) to (u_1, u_2) obtained by rotating the axis by an angle α.

express the transformation as a matrix operation. Let us define the basis as a matrix (with each vector v_1 and v_2 as column vectors).

$$V = \begin{pmatrix} 1 & 0 \\ 0 & 1 \end{pmatrix} \tag{4.86}$$

we can define the transformation matrix T as

$$T = \begin{pmatrix} \cos\alpha & -\sin\alpha \\ \sin\alpha & \cos\alpha \end{pmatrix} \tag{4.87}$$

Then the new basis matrix U (where we have indicated with $u_{1,x}$ the component of the vector u_1 along v_1, etc.) with the new basis vectors as columns

$$U = \begin{pmatrix} u_{1,x} & u_{2,x} \\ u_{1,y} & u_{2,y} \end{pmatrix} \tag{4.88}$$

can be calculated with

$$U = TV \tag{4.89}$$

It is easy to check that this is correct. Let's now consider a point $P = p_1 v_1 + p_2 v_2$ (we will consider the basis vectors in matrix formalism, as column vectors), that can be written in vector form as

$$P = p_1 v_1 + p_2 v_2 = p_1 \begin{pmatrix} 1 \\ 0 \end{pmatrix} + p_2 \begin{pmatrix} 0 \\ 1 \end{pmatrix} = \begin{pmatrix} p_1 \\ p_2 \end{pmatrix} \tag{4.90}$$

How can we express P in the new basis (which we will indicate with P_V)? This can now be easily done using the transformation matrix T. In fact we have

$$P_V = TP = \begin{pmatrix} \cos\alpha & -\sin\alpha \\ \sin\alpha & \cos\alpha \end{pmatrix} \begin{pmatrix} p_1 \\ p_2 \end{pmatrix} = \begin{pmatrix} p_1\cos\alpha - p_2\sin\alpha \\ p_1\sin\alpha + p_2\cos\alpha \end{pmatrix} \tag{4.91}$$

This can be easily verified by drawing a diagram similar to Figure 4.9. In general, given a dataset X where each column is a single sample of our data, when we have a linear transformation T, the new representation Y in the new basis will be given by

$$Y = TX \tag{4.92}$$

that is the natural expansion of Equation 4.91. In fact, in Equation 4.91 we just calculated the new coordinates of a single point, in Equation 4.92 we changed the coordinates of **all** points at the same time (since each is a column in the matrix X).

Equation 4.91 represents a change of basis. In general, T is a rotation and a stretch of space. The rows of T represent the new basis vectors. Note that the matrix T can have a higher dimension that 2×2 as in our example. In fact, X is our dataset, and that means that the vertical dimension is given by the number of features that we have and that can be very high. This can be easily seen by writing the equation in the following form (assuming that the matrix T has m rows and our dataset has n data samples, or the number of columns).

$$TX = \begin{pmatrix} - \boldsymbol{t}_1 - \\ \vdots \\ - \boldsymbol{t}_m - \end{pmatrix} \begin{pmatrix} | & & | \\ \boldsymbol{x}_1 & \cdots & \boldsymbol{x}_n \\ | & & | \end{pmatrix} = \begin{pmatrix} \boldsymbol{t}_1 \cdot \boldsymbol{x}_1 & \cdots & \boldsymbol{t}_1 \cdot \boldsymbol{x}_n \\ \vdots & \ddots & \vdots \\ \boldsymbol{t}_m \cdot \boldsymbol{x}_1 & \cdots & \boldsymbol{t}_m \cdot \boldsymbol{x}_n \end{pmatrix} \tag{4.93}$$

And as you can see, the first row of the matrix TX will be the projection of the the dataset along the vector $\boldsymbol{t}_1$, the second along $\boldsymbol{t}_2$ and so on.

In general, any linear transformation can be interpreted, in geometric terms, as a rotation, a stretch, or some other geometrical modification [12] (for example reflection with respect to some line). In Table 4.1 a list of possible transformations [12] and the respective transformation matrices in two dimensions is reported.

Geometrical Transformation	**Transformation Matrix**
Rotation of an angle α	$\begin{pmatrix} \cos\alpha & -\sin\alpha \\ \sin\alpha & \cos\alpha \end{pmatrix}$
Reflection through the x-axis	$\begin{pmatrix} 1 & 0 \\ 0 & -1 \end{pmatrix}$
Reflection through the y-axis	$\begin{pmatrix} -1 & 0 \\ 0 & 1 \end{pmatrix}$
Reflection through the origin	$\begin{pmatrix} 0 & 1 \\ 1 & 0 \end{pmatrix}$

Table 4.1 List of transformation matrices for specific geometrical modifications [12].

4.6.4 PCA Formalisation

Now that we have the formalism out of the way, let us go back to the original question. With this formalism, we still have to answer two main questions.

- What is the best way to rewrite our dataset X? Or in other words, how can we decide what is a good and what is a bad representation (a good or bad basis)?
- What is a good choice for T?

To answer the first question, PCA relies on two key assumptions:

1. The directions along which the data show the largest variance are the ones that contain the most relevant and interesting information.
2. Features that are correlated to each other are *redundant*, or in other words not useful in describing the phenomena we are studying.

Hypothesis 1 can be grasped by considering that if there is minimal variance along a certain direction, it implies that the data remain largely unchanged in that direction, suggesting a lack of significant information. Take, for instance, our spring example: the direction orthogonal to the spring extension bears little relevance to describing the phenomenon. In this case, the movement of objects attached to the spring in the perpendicular direction would be negligible (or even non-existent), resulting in very low variance along this axis. Therefore, our objective is to identify a new basis on which the variance is maximised along some (or all) of its directions.

To summarise again, our goal is twofold: first, we need to identify the directions along which the variance is largest, and second, we need to identify any redundant features (so that we can safely ignore them). In the next section, I will describe one way of achieving this.

4.6.5 Covariance Matrix

One problem that we have not discussed is how to determine if one variable is redundant in relation to another. This is essentially about assessing whether the two variables are correlated. For example, consider a dataset with two features represented as $D = \{x_i, x_i\}_{i=1}^N$. It is clear that in this case, we don't require both features to describe D, as they are identical. One feature would certainly be adequate, meaning that the second is *redundant*.

To explain this in more general terms, let us consider two sets of measurements: $X = \{x_i\}_{i=1}^N$ and $Y = \{y_i\}_{i=1}^N$. To simplify the equations, let us suppose that both have zero mean. In this case, the *covariance* of X and Y

$$\sigma_{XY}^2 = \frac{1}{N}\sum_{i=1}^{N} x_i y_i \tag{4.94}$$

measures the degree of relationship between the two variables. A large positive value indicates positively correlated data (if one grows, so does the other). A negative value indicates negatively correlated data (if one grows, the other decreases). By writing X and Y as matrices $\boldsymbol{X}$ and $\boldsymbol{Y}$ with dimensions $(1, N)$ we can rewrite the covariance as a matrix product

$$\sigma_{XY}^2 = \frac{1}{N}\boldsymbol{X}\boldsymbol{Y}^T \tag{4.95}$$

In general if we have a large dataset we can generalize the definition. Consider a data set $\boldsymbol{X}$

$$\boldsymbol{X} = \begin{pmatrix} \text{—} \boldsymbol{x}_1 \text{—} \\ \vdots \\ \text{—} \boldsymbol{x}_m \text{—} \end{pmatrix} \tag{4.96}$$

where each row contains a complete measurement (all the features), while each column contains all measurements of a specific feature[8]. In this case we can write the **covariance matrix** $\boldsymbol{C}$ as

$$\boldsymbol{C} = \frac{1}{N}\boldsymbol{X}\boldsymbol{X}^T \tag{4.97}$$

Note that $\boldsymbol{C}$ is a square matrix, its diagonal terms are the variances of the features and the off-diagonal terms the covariance between pairs of different features.

Remembering our hypothesis, that what is interesting happens along directions that have large variance and small co-variance (low redundancy), we can make the following statements:

1. Large diagonal elements indicate interesting features.
2. large off-diagonal terms indicate large redundancy.

You may at this point see where we are going with this. Our goal is to **find a new basis (diagonalise the covariance matrix) to (1) maximise the variance (so the diagonal elements) and (2) minimise redundancy (the off-diagonal elements)**. This is accomplished by diagonalising the covariance matrix, which is essentially the entirety of what PCA entails.

To make PCA easy to use and calculate, PCA assumes that the new basis will be an **orthonormal matrix**, or in other words, that the vectors of the new basis are orthogonal to each other.

4.6.6 Overview of Assumptions

Let us review all the assumptions PCA makes.

1. *Linearity*: the change of basis we are searching for is a linear transformation (there are other approaches that lift this assumption, like t-SNE or t-distributed Stochastic Neighbour Embedding [13]).
2. *Large variance is important*: this assumption is often safe to make, but it is a strong assumption that is not always correct. For example, if you are studying the transverse oscillations of a spring due to imperfections in your system, PCA may simply ignore those effects since the variance is minimal along the transverse direction.
3. *The new basis is orthogonal*: this makes using PCA fast and easy, but it does not work every time. It may well be that there are directions where the interesting phenomena occur that are not orthogonal to each other.

[8] This is done in the opposite way as before, since in this case we are interested in the covariance between features and not measurements.

4.6.7 PCA with Eigenvectors and Eigenvalues

The problem that PCA solves can be stated as follows.

Problem 4.1 (PCA Problem Statement) Find an orthonormal matrix $\boldsymbol{T}$, such that, given a dataset $\boldsymbol{X}$ and $\boldsymbol{Y} = \boldsymbol{T}\boldsymbol{X}$ the matrix $\boldsymbol{C} = (1/N)\boldsymbol{Y}\boldsymbol{Y}^T$ is a diagonal matrix.

Let us start by writing $\boldsymbol{C}$.

$$\begin{aligned}
\boldsymbol{C} &= \frac{1}{N}\boldsymbol{Y}\boldsymbol{Y}^T \\
&= \frac{1}{N}(\boldsymbol{T}\boldsymbol{X})(\boldsymbol{T}\boldsymbol{X})^T \\
&= \frac{1}{N}\boldsymbol{T}\boldsymbol{X}\boldsymbol{X}^{\boldsymbol{T}}\boldsymbol{T}\boldsymbol{T} \\
&= \boldsymbol{T}\left(\frac{1}{N}\boldsymbol{X}\boldsymbol{X}^T\right)\boldsymbol{T}^T \\
&= \boldsymbol{T}\boldsymbol{C}_{\boldsymbol{X}}\boldsymbol{T}^T
\end{aligned} \tag{4.98}$$

where with $\boldsymbol{C}_{\boldsymbol{X}}$ we have indicated the covariance matrix of $\boldsymbol{X}$. Now you should know that any symmetric matrix $\mathbf{M}$ is diagonalized by a matrix composed of its eigenvectors organized as columns. The trick now is to choose $\boldsymbol{T}$ to be a matrix where each row is an eigenvector of $\left(\frac{1}{N}\boldsymbol{X}\boldsymbol{X}^T\right)$. Let us indicate this matrix with $\boldsymbol{E}^T$, and let us indicate with $\boldsymbol{D}$ the diagonalised version of $\left(\frac{1}{N}\boldsymbol{X}\boldsymbol{X}^T\right)$.

$$\begin{aligned}
\boldsymbol{C} &= \boldsymbol{T}\boldsymbol{C}_{\boldsymbol{X}}\boldsymbol{T}^T \\
&= \boldsymbol{T}(\boldsymbol{E}\boldsymbol{D}\boldsymbol{E}^T)\boldsymbol{T}^T \\
&= \{\text{Remember } \boldsymbol{E}^T = \boldsymbol{T}\} \\
&= (\boldsymbol{T}\boldsymbol{T}^T)\boldsymbol{D}(\boldsymbol{T}\boldsymbol{T}^T) \\
&= \{\text{Since the matrix } \boldsymbol{T} \text{ is orthonormal then } \boldsymbol{T}^T = \boldsymbol{T}^{-1}\} \\
&= (\boldsymbol{T}\boldsymbol{T}^{-1})\boldsymbol{D}(\boldsymbol{T}\boldsymbol{T}^{-1}) \\
&= \boldsymbol{D}
\end{aligned} \tag{4.99}$$

So this choice of $\boldsymbol{T}$ diagonalizes $\boldsymbol{C}$. This is how to calculate the transformation matrix $\boldsymbol{T}$, simply by calculating the eigenvectors of the covariance matrix of our data $\boldsymbol{X}$. There is actually another way of finding $\boldsymbol{T}$, and that is by using singular value decomposition (SVD), but that goes beyond the scope of this book. This is what Python and R libraries typically use, and they typically return the dataset in the new basis, already prepared for you.

Therefore, finding the eigenvalues and eigenvectors allows us to find a basis transformation that diagonalizes the covariance matrix of $\boldsymbol{X}$, thus maximising the diagonal elements (the directions where something interesting happens) and minimising the off-diagonal elements (eliminating the redundancies in the data).

4.6.8 One Implementation Limitation

All we discussed sound good, but one limitation of PCA is that when you want to numerically diagonalise the matrix $\left(\frac{1}{N}\boldsymbol{X}\boldsymbol{X}^T\right)$, this must fit completely in memory. With a large dataset, this may be a problem (often is). This is why you may encounter difficulties in doing this. One variant of PCA that you may consider is IPCA (Incremental Principal Component), that goes beyond the scope of this book, but is available in the Python library scikt-learn [14].

References

1. Adam Paszke, Sam Gross, Francisco Massa, Adam Lerer, James Bradbury, Gregory Chanan, Trevor Killeen, Zeming Lin, Natalia Gimelshein, Luca Antiga, Alban Desmaison, Andreas Kopf, Edward Yang, Zachary DeVito, Martin Raison, Alykhan Tejani, Sasank Chilamkurthy, Benoit Steiner, Lu Fang, Junjie Bai, and Soumith Chintala. Pytorch: An imperative style, high-performance deep learning library. In H. Wallach, H. Larochelle, A. Beygelzimer, F. d'Alché-Buc, E. Fox, and R. Garnett, editors, *Advances in Neural Information Processing Systems 32*, pages 8024–8035. Curran Associates, Inc., 2019.
2. Martín Abadi, Ashish Agarwal, Paul Barham, Eugene Brevdo, Zhifeng Chen, Craig Citro, Greg S. Corrado, Andy Davis, Jeffrey Dean, Matthieu Devin, Sanjay Ghemawat, Ian Goodfellow, Andrew Harp, Geoffrey Irving, Michael Isard, Yangqing Jia, Rafal Jozefowicz, Lukasz Kaiser, Manjunath Kudlur, Josh Levenberg, Dandelion Mané, Rajat Monga, Sherry Moore, Derek Murray, Chris Olah, Mike Schuster, Jonathon Shlens, Benoit Steiner, Ilya Sutskever, Kunal Talwar, Paul Tucker, Vincent Vanhoucke, Vijay Vasudevan, Fernanda Viégas, Oriol Vinyals, Pete Warden, Martin Wattenberg, Martin Wicke, Yuan Yu, and Xiaoqiang Zheng. TensorFlow: Large-scale machine learning on heterogeneous systems, 2015. Software available from tensorflow.org.
3. Lloyd N Trefethen and David Bau III. *Numerical linear algebra*, volume 50. Siam, 1997.
4. Peter D. Lax. *Linear Algebra and Its Applications*. Wiley-Interscience, Hoboken, NJ, second edition, 2007.
5. Harry Dym. *Linear algebra in action*, volume 78. American Mathematical Soc., 2013.
6. Charles R. Harris, K. Jarrod Millman, Stéfan J. van der Walt, Ralf Gommers, Pauli Virtanen, David Cournapeau, Eric Wieser, Julian Taylor, Sebastian Berg, Nathaniel J. Smith, Robert Kern, Matti Picus, Stephan Hoyer, Marten H. van Kerkwijk, Matthew Brett, Allan Haldane, Jaime Fernández del Río, Mark Wiebe, Pearu Peterson, Pierre Gérard-Marchant, Kevin Sheppard, Tyler Reddy, Warren Weckesser, Hameer Abbasi, Christoph Gohlke, and Travis E. Oliphant. Array programming with NumPy. *Nature*, 585(7825):357–362, September 2020.
7. Kaare Brandt Petersen, Michael Syskind Pedersen, et al. The matrix cookbook. *Technical University of Denmark*, 7(15):510, 2008.
8. Sparsh Mittal and Shraiysh Vaishay. A survey of techniques for optimizing deep learning on gpus. *Journal of Systems Architecture*, 99:101635, 2019.
9. Norman P Jouppi, Cliff Young, Nishant Patil, David Patterson, Gaurav Agrawal, Raminder Bajwa, Sarah Bates, Suresh Bhatia, Nan Boden, Al Borchers, et al. In-datacenter performance

analysis of a tensor processing unit. In *Proceedings of the 44th annual international symposium on computer architecture*, pages 1–12, 2017.
10. Amna Shahid and Malaika Mushtaq. A survey comparing specialized hardware and evolution in tpus for neural networks. In *2020 IEEE 23rd International Multitopic Conference (INMIC)*, pages 1–6. IEEE, 2020.
11. Lindsay I Smith. A tutorial on Principal Components Analysis.
12. John Gilbert. Linear Transformations. https://web.ma.utexas.edu/users/gilbert/. [Last accessed 1st Oct. 2023].
13. Laurens Van der Maaten and Geoffrey Hinton. Visualizing data using t-sne. *Journal of machine learning research*, 9(11), 2008.
14. Incremental PCA. `https://scikit-learn/stable/auto_examples/decomposition/plot_incremental_pca.html`. [Last accessed 1st Oct. 2023].

Chapter 5
Statistics and Probability for Machine Learning

You can, for example, never foretell what any one man will do, but you can say with precision what an average number will be up to. Individuals vary, but percentages remain constant. So says the statisticians.

Sherlock Holmes, The Sign of Four

This chapter delves into the critical role of statistics and probability in machine learning, starting with an overview of random experiments and variables. It progresses to cover essential topics such as set theory, probability, conditional probability, and key theorems such as the Bayes Theorem and the Central Limit Theorem. The discussion extends to distribution functions, the significance of expected values and variance, and the normal distribution, along with a variety of other distributions relevant to machine learning. The chapter also introduces Moment Generating Functions as vital tools in probability theory, providing a foundation for understanding the Central Limit Theorem's implications for data analysis and prediction in machine learning. By exploring these statistical concepts, the chapter aims to equip readers with the necessary knowledge to effectively engage with machine learning models and understand the statistical underpinnings of algorithm performance and data analysis.

5.1 Motivation

Statistics forms the foundation of nearly all machine learning algorithms, including neural networks. Its primary goal is to infer characteristics of a larger population based on experimental observations from a subset of that population.

Machine learning seeks to identify patterns within a sample of data, thus inferring broader insights about the entire population from this limited dataset. Here, the term "**population**" refers to the complete set of data points (or measurements) relevant to a specific research question (for a detailed discussion of what a research question is, see Chapter 6), while a "**sample**" is a representative segment of that population. Access to the entire population is rare; for instance, in studying patients with heart conditions, it is impossible to gather data on every individual worldwide with such conditions at any given moment. Therefore, we usually rely on a sample of the population. Statistics play a crucial role in this context, allowing us to derive meaningful conclusions about the entire population based on sample data. When

U. Michelucci, *Fundamental Mathematical Concepts for Machine Learning in Science*,
https://doi.org/10.1007/978-3-031-56431-4_5

training machine learning models, the objective is to ensure their applicability to the entire population, despite being trained only on a (typically) much smaller sample.

The primary challenge in machine learning, often referred to as the challenge of generalization, is devising models that can learn universal traits applicable to the entire population using just a sample dataset. Machine Learning methodologies are deeply intertwined with statistical principles. Some methods, such as generative learning and specific neural network architectures like variational autoencoders or Generative Adversarial Networks, demand an in-depth understanding of statistics. Bayes Neural Networks, for example, fall into this category, though they are not covered in this book due to their complexity. Furthermore, statistics are instrumental in evaluating how algorithms perform on new data, validating and comparing different models, tuning hyper-parameters, devising efficient resampling methods for small datasets, and more. Proficiency in statistics is an indispensable skill for any data scientist.

5.2 Random Experiments and Variables

In statistics we consider experiments for which the outcome cannot be predicted with certainty, those are called **random experiments**.

Definition 5.1 (Random Experiment) Experiments for which the outcome cannot be predicted with certainty are called **random experiments**.

As examples of random experiments you can consider are

- tossing of a coin;
- tossing a dice;
- measuring the height of a person randomly chosen from a certain city.

In all the examples you cannot predict what the outcome of the experiment will be, therefore those would be classified as random experiments.

Info Random Experiments and Neural Networks

This may seem irrelevant in the context of machine learning. But remember that the output of a model cannot be predicted exactly. For example, by training a network two times with a different random seed for the numerical weight initialisation will lead to different weights, and a slightly different output. Simply splitting your dataset in a different way (in Chapter 7 you will learn that one split a dataset in training and validation) will also lead to a different model, different weights, and thus different output. This means that one can consider, under certain assumptions, the prediction of a certain output of a neural network as a random experiment. Thus statistics will play a huge role in discussing it, as we will see in more detail later on.

To move forward, further definitions are required to ensure the accuracy of our discussion on probability.

Definition 5.2 (Outcome Space S**)** The collection of all possible outcomes of a random experiment is called the **outcome space** and is indicated with S.

Having defined the outcome space, we need to define what we mean by an event. In this section, we will use some very basic set notation. In case you are unsure of the meaning, you can quickly check the first paragraphs of Section 5.3.

Definition 5.3 (Event A**)** An **event** A is a subset[1] of the outcome space S: $A \subset S$.

Example Outcome Space and Events

To be a bit clearer, let us consider an example. Suppose you have a six-face fair dice. The random experiment we consider will be the throwing of the dice together with the observation of the number that is on the top face when the dice come to rest on the surface you have thrown the dice on. Now, since the dice will not disappear mid-air, one of the 6 numbers must come up, thus the outcome space will be $S = \{1, 2, 3, 4, 5, 6\}$. For example, the event of getting a 6 would be $A = \{6\}$, or the event of getting 2 will be $A = \{2\}$. Note that all the events described are subsets of S. An event cannot contain something that is not in the outcome space. In our example, $\{7\}$ is not an event as will never happen since the dice has only six faces. An event can also contain multiple elements, depending on the random experiment. Consider, for example, the experiment of throwing two dice at the same time. The outcome space will now be larger

$$S = \{(1, 1), (1, 2), ..., (1, 6), (2, 1), ..., (2, 6), ..., (6, 6)\} \tag{5.1}$$

and will contain 36 elements. Now, the event of getting the same number on both dices will be a set of six elements

$$A = \{(1, 1), (2, 2), (3, 3), (4, 4), (5, 5), (6, 6)\} \tag{5.2}$$

It is important to note that events and outcome spaces depend on what random experiment you are performing (as we have seen in this example).

The second concept that is the basis of statistics is that of *random variable*.

Definition 5.4 (Random Variable) A **random variable** X is a function from a sample space S on the real numbers $X : S \to \mathbb{R}$.

[1] A subset is a portion of a given set of elements. For example, if you have the set of integers from 1 to 10 $\{1, 2, 3, 4, 5, 6, 7, 8, 9, 10\}$, examples of subsets of it could be the even $\{2, 4, 6, 8, 10\}$ or the odd integers of it $\{1, 3, 5, 7, 9\}$.

Example **Random Variables**

In our example of tossing two dice, a random variable could be the sum of the numbers coming out for each roll. Or, if you consider tossing just one dice, a random variable could be the sum of the results of 50 tosses. Usually, a random variable is indicated with a big letter like X or Y.
A random variable is called "**variable**" because it represents a value that can vary due to chance. The term "variable" emphasises that, unlike a constant, the value it takes is not fixed but can change depending on the outcome of the random process with which it is associated.

Many theorems in probability are proved by using sets. The proofs are not mandatory, but it is nonetheless useful to know the main idea behind them. The next section will very briefly summarise some properties of sets and their operations that you can use as a reference.

5.3 Algebra of Sets

As it should be clear from the definitions in the previous section, to study probability the language of sets is widely used. Therefore is a good idea to review some notation that we will use in the following sections.

- Empty set: $\varnothing$.
- A is a subset of B: $A \subset B$. Basically this means that all elements in A are also in B.
- Union of A and B: $A \cup B$. This means that the set $A \cup B$ contains all elements that are in A **or** in B.
- Intersection of A and B: $A \cap B$. This means that the set $A \cap B$ contains all elements that are A **and** in B.
- A' is the complement of A. This means that A' contains all elements available that are **not** in A.
- The symbol $\bigcup$ (union of multiple sets) must be interpreted in this way (formula given for a finite number of elements, but that remains valid also for an infinite number)

$$\bigcup_{i=1}^{N} A_i = A_1 \cup A_2 \cup A_3 \cup \ldots \cup A_N \tag{5.3}$$

  or in words, the set of elements that are in **at least** one of the A_i for $i = 1, \ldots, N$.
- The symbol $\bigcap$ (intersection of multiple sets) must be interpreted in this way (formula given for a finite number of elements, but that remains valid also for an infinite number)

$$\bigcap_{i=1}^{N} A_i = A_1 \cap A_2 \cap A_3 \cap \ldots \cap A_N \tag{5.4}$$

or in words, the set of elements that are in **all** the A_i for $i = 1, ..., N$.

Thanks to this notation we can also give a definition of mutually exclusive events and exhaustive events.

Definition 5.5 (Mutually exclusive events) Given an outcome space S, we call **mutually exclusive events** A_i with $i = 1, ..., k$ (we consider here only a finite set of events) when $A_i \cap A_j = \emptyset$, for $i \neq j$.

Definition 5.6 (Exhaustive events) Given an outcome space S, we call **exhaustive events** A_i with $i = 1, ..., k$ (we consider here only a finite set of events) when $\bigcup_{i=1}^{k} A_i = S$.

Example Mutually Exclusive and Exhaustive Events

In our example of tossing a dice, the events $A_i = \{i\}$ for $i = 1, ..., 6$ are clearly mutually exclusive events and at the same time also exhaustive events, as it is easy to verify. In our second example of tossing two dices at the same time, the events of getting the same numbers on the two dices defined by $A_i = \{(i, i)\}$ for $i = 1, ..., 6$ are clearly mutually exclusive events, but surely not exhaustive events.

Since we are discussing sets, it is useful to give the three most important laws of how sets can be composed. The most important set laws are the following. Suppose A, B and C are sets. Then the following are valid.

- **Associative** - $(A \cup B) \cup C = A \cup (B \cup C)$ - $(A \cap B) \cap C = A \cap (B \cap C)$
- **Distributive** - $A \cap (B \cup C) = (A \cap B) \cup (A \cap C)$ - $A \cup (B \cap C) = (A \cup B) \cap (A \cup C)$
- **De Morgan's Law** - $(A \cup B)' = A' \cap B'$ - $(A \cap B)' = A' \cup B'$

These will be useful in proving some of the main theorems about probability that we will discuss later. The proofs are not difficult but goes beyond the introductory nature of this chapter.

5.4 Probability

Now we have all the instruments to start discussing what probability is and how to define it. We will start with the most intuitive and practical definition: the relative frequency interpretation.

5.4.1 Relative Frequency Interpretation of Probability

Suppose we repeat an experiment n times and we count the number of times we observe event A as output. We indicate this number with $N(A)$. Note that naturally, even if not explicitly indicated, $N(A)$ is dependent on n.

Definition 5.7 (Relative Frequency of an Event) We define the **relative Frequency of event** A as

$$\frac{N(A)}{n} \tag{5.5}$$

Note that this is rather rapidly varying numbers in practice, but that it tends to vary less as n grows. We can define

$$p = \lim_{n \to \infty} \frac{N(A)}{n} \tag{5.6}$$

and we will call p the probability of event A.

Definition 5.8 (Relative Frequency Definition of Probability) We define the probability of an event with the limit for $n \to \infty$ of $N(A)/n$.

It is useful to define what the term "prior probability" means, as it is often used.

Definition 5.9 (Prior Probability) Probability that is known before doing any experiments. This can be evaluated if we know S and the occurrence frequencies of each element.

Example Prior Probability when Throwing a Dice

Consider again our example of the dice with six faces that is thrown. We have already discussed that the outcome space is $S = \{1, 2, 3, 4, 5, 6\}$. Suppose we know it is a fair dice, and therefore each face will have the same probability of occurrence. Then we can estimate the prior probability of one face as output as $1/6$.

5.4.2 Probability as a Set Function

One of the most general ways of defining probability is to define it as a set function.

Definition 5.10 (Probability as a set function) In general we can define probability as a **set function** P that associates with any subset of S (in other words with an event A) a real number $p \in [0, 1]$. The probability of an event A is indicated with $P(A)$.

5.4.3 ★ Axiomatic Definition of Probability

The previous discussion about probability is intuitive and will serve the student well since it is easy to interpret and apply to a practical case. But from a mathematical point of view, it misses the necessary rigour that will allow us to derive many important results in statistics. To give an **axiomatic** definition of probability, we first need to define what a sigma algebra [1] is.

Definition 5.11 A collection of subsets of S A_i is called a sigma algebra denoted by $\mathcal{B}$, if it satisfies the following three properties:

1. $\varnothing \in \mathcal{B}$ (the empty set is an element of $\mathcal{B}$)
2. If $A \in \mathcal{B}$, then $A' \in \mathcal{B}$ ($\mathcal{B}$ is closed under complementation)
3. If $A_1, A_2, ... \in \mathcal{B}$ then $\cup_{i=1}^{\infty} A_i \in \mathcal{B}$ ($\mathcal{B}$ is closed under countable unions)

Let us note a few things here. Property 1. states that since $\varnothing \subset S$, and since $S = \varnothing'$ then also S is in $\mathcal{B}$. Then we can also note from property 3 and the DeMorgan's law we have

$$\left(\bigcup_{i=1}^{\infty} A_i\right)' = \bigcap_{i=1}^{\infty} A_i \tag{5.7}$$

and therefore for property 2 of the algebra, $\bigcap_{i=1}^{\infty} A_i \in \mathcal{B}$.

Example Sigma Algebra

Given an outcome space S (finite or countable) the simplest example of a sigma algebra we can define is

$$\mathcal{B} = \{\text{all subsets of } S \text{ including } S \text{ itself}\} \tag{5.8}$$

The following theorem gives the number of elements that this sigma algebra has.

Theorem 5.1 *Is S has n elements, there are* 2^n *subsets in* $\mathcal{B}$.

Proof This theorem can be easily proven by considering that:

- There are n sets with just 1 element.
- There are $n(n-1)/2$ sets with 2 elements. Note that the 2 at the denominator is there, since we are not interested in the order of the elements. Therefore for example 1,2 and 2,1 are equivalent. Recall that, in general, there are $k!$ ways of ordering k elements.
- There are $n(n-1)(n-2)/3!$ sets with 3 elements.
- And so on.

In general there are $n!/(n-k)!k!$ sets with k elements in S. Note that the expression is nothing else than the binomial coefficient

$$\binom{n}{k} = \frac{n!}{(n-k)!k!} \tag{5.9}$$

Therefore, the total number of sets in S can be written as (the 1 in front is there to count also the empty set):

$$1 + \binom{n}{1} + \binom{n}{2} + \ldots + \binom{n}{n} = \sum_{i=0}^{n} \binom{n}{i} = 2^n \tag{5.10}$$

is easy to see that the sum of binomials is 2^n since we can use the well-known formula for the expansion of $(x + y)^n$

$$(x + y)^n = \sum_{i=0}^{n} \binom{n}{k} x^{n-k} y^k \tag{5.11}$$

and for our case it is sufficient to choose $x = 1$ and $y = 1$. □

Definition 5.12 Given a sample space S and an associated sigma algebra $\mathcal{B}$, a probability function $P(A)$ with domain $\mathcal{B}$ is a real-valued set function $P : \mathcal{B} \to \mathbb{R}$ such that

1. $P(A) \geq 0$ for all $A \in \mathcal{B}$
2. $P(S) = 1$
3. if $A_i \in \mathcal{B}$ and $A_i \cap A_j = \varnothing$ for $i \neq j$ (mutually exclusive events) then $P(\cup_{i=1}^{\infty} A_i) = \sum_{i=1}^{\infty} P(A_i)$

Note that the three properties given in the definition are referred as the **axioms of probability** or the **Kolmogorov Axioms**. Any function that satisfies the axioms is called a probability function. But there are no guidelines in the axioms on how to choose P. For any sample space (and algebra) one can choose different probability functions, as long as they satisfy the axioms.

We can now give a definition of independent events.

Definition 5.13 Given two events A and B, they are said to be **independent** if and only if

$$P(A \cap B) = P(A)P(B)$$

5.4.4 Properties of Probability Functions

Of course, ignoring the axioms we discussed to determine whether a function P truly qualifies as a probability function is impractical. But there is a theorem that can help us in defining probability functions that automatically satisfy the Kolmogorov axioms discussed in the previous section. This will also explain, formally, why the softmax function is a probability function (more on that later).

Theorem 5.2 *Let $S = \{s_1, ..., s_n\}$ be a finite set. Let $\mathcal{B}$ be any sigma algebra of subsets of S. Let $p_1, ..., p_n$ positive real numbers such that $\sum_{i=1}^{n} p_i = 1$. For any $A \in \mathcal{B}$, define $P(A)$ by*

$$P(A) = \sum_{i:s_i \in A} p_i \tag{5.12}$$

Then P is a probability function on $\mathcal{B}$. This remains true if S is a countable set.

Proof Let consider S finite for simplicity. First of all since all $p_i \geq 0$ then it is true that for any $A \in \mathcal{B}$, $P(A) = \sum_{i:s_i \in A} p_i \geq 0$. Thus Axiom 1 is true. By how we choose the p_i is also true that

$$P(S) = \sum_{i:s_i \in S} p_i = \sum_{i=1}^{n} p_i = 1 \tag{5.13}$$

thus also Axiom 2 is true. Now let us consider k disjoint events A_i, $i = 1, ...k$. Then

$$P\left(\bigcup_{i=1}^{k} A_i\right) = \sum_{j:s_j \in \cup_{i=1}^{k} A_i} p_j = \sum_{i=1}^{k} \sum_{j:s_j \in A_i} p_j = \sum_{i=1}^{k} P(A_j) \tag{5.14}$$

Thus also axiom 3 is verified and P is a probability function. □

5.5 The Softmax Function

The softmax function is a crucial tool in machine learning, particularly in the context of classification problems. It takes a vector of numbers and transforms them into a probability distribution, where the probabilities of each class are proportional to the exponential of the input numbers. The key feature of softmax is that it accentuates the differences in the input values, making the highest values stand out more and the lower ones fade, which is particularly useful in multi-class classification scenarios. In practical terms, the softmax function is often used in the final layer of a neural network for tasks like image recognition or language processing, where the goal is to categorize inputs into multiple classes. The output probabilities make it easier to determine which class the model predicts as the most likely.

Let us formulate it in a more mathematical form. Given a vector $\mathbf{v} = (v_1, ..., n_n)$ with n some integer the softmax function is a function $\mathrm{SM} : \mathbb{R}^n \rightarrow \mathbb{R}^n$ that gives as output for its ith component

$$\mathrm{SM}(\mathbf{v})_i = \frac{e^{v_i}}{\sum_{i=1}^{n} e^{v_i}} \tag{5.15}$$

you can easily see that

$$\sum_{i=1}^{n} \mathrm{SM}(\mathbf{v})_i = 1 \tag{5.16}$$

and thus this can be interpreted as a probability. In a classification problem, $\mathrm{SM}(\mathbf{v})_i$ is interpreted as the probability of the input in being in class i. Normally to choose a single class the component with the highest probability will be chosen. For example if you have $\mathrm{SM}(\mathbf{v}) = (0.2, 0.4, 0.1, 0.1, 0.1, 0.1)$ (suppose we have 6 components), then you would classify your input $\mathbf{v}$ in class number 2 (the one with a 0.4 probability).

Info ★ Softmax is a Probability Function that Satisfy the Kolmogorov Axioms

From Theorem 5.2 it is easy to see that the softmax is a probability function. In fact, let us consider the case where we want to classify an observation with a neural network of some kind in k classes. Typically, the network will have an output layer with k neurons having as activation function the softmax function. That means that we can write $S = \{1, 2, ..., k\}$, thus we have a finite S. The softmax function simply gives as output k values that are positive and sum to 1. This is exactly the definition given in Theorem 5.2 and therefore the softmax function is a probability function that satisfies the Kolmogorov axioms.

5.5.1 ■ Softmax Range of Applications

The softmax function is widely used in various machine learning scenarios, extending beyond the realm of neural networks. Here are some examples where the softmax function is useful:

1. **Multinomial Logistic Regression:** in logistic regression models dealing with multi-class classification problems, the softmax function is used to generalize the binary logistic regression to multiple classes. Instead of determining the probability of a single class, softmax is employed to calculate the probabilities of each class over all possible classes.
2. **Decision Tree Ensembles:** in ensemble methods such as Gradient Boosting or Random Forests, which involve a combination of multiple decision trees, the softmax function can be used to aggregate the predictions from various trees. This approach helps in providing a probabilistic interpretation of multi-class predictions.
3. **Genetic Algorithms:** in certain implementations of genetic algorithms for optimization problems with multiple potential solutions, softmax can be utilized to probabilistically select solutions based on their fitness scores. This helps in maintaining diversity in the population.
4. **Markov Decision Processes (MDPs):** in the field of reinforcement learning, the softmax function can be used in policy representation. It provides a method to convert values or scores associated with different actions into action probabilities, aiding in decision-making processes.
5. **K-Nearest Neighbors (KNN):** softmax can be used in KNN for multi-class classification tasks. After determining the K-nearest neighbors, softmax can assign probabilities to each class based on the distribution of neighbors, providing a more nuanced classification than a simple majority vote.
6. **Probabilistic Graphical Models:** in graphical models such as Hidden Markov Models (HMMs) or Conditional Random Fields (CRFs), the softmax function

is often used in the normalization step to compute the probabilities of different states or sequences, ensuring that they sum to one.
7. **Clustering Algorithms:** in clustering scenarios, particularly in methods like Soft Clustering where data points can belong to multiple clusters with different degrees of membership, softmax can be used to assign probabilities to each cluster for a given data point.
8. **Dimensionality Reduction Techniques:** in techniques like t-Distributed Stochastic Neighbor Embedding (t-SNE), softmax is used in the computation of pairwise affinities between points, helping to maintain the structure of the data in reduced dimensions.
9. **Game Theory and Multi-Agent Systems:** in multi-agent systems and game theory, softmax is sometimes used for strategy selection based on payoff or utility values. Agents can use softmax to probabilistically choose actions, balancing exploration and exploitation.

Softmax is widely used function that any researcher should know, as it is a necessary tool in a wide variety of scenarios.

5.6 ★ Some Theorems about Probability Functions

In this section we list some theorems that highlight properties of probability functions.

Theorem 5.3 $P(A) = 1 - P(A')$

Proof We can write $S = A \cup A'$ and we know that $A \cap A' = \varnothing$ (by definition). Therefore

$$1 = P(S) = P(A \cup A') = P(A) + P(A')$$

from the third property of probability functions (see Definition 5.12). □

Theorem 5.4 $P(\varnothing) = 0$

Proof This is the intuitive result that in any experiment we **must** have at least some output. If you roll a dice, you cannot have no result. This can be proven mathematically with the definition of the probability function and the algebra of sets. In fact, we can define $A = \varnothing$ and $A' = S$. Then for Theorem 5.3 we have

$$P(\varnothing) = 1 - P(S) = 0$$

Theorem 5.5 *If* $A \subset B$ *then* $P(A) \leq P(B)$

Proof We can write

$$B = A \cup (B \cap A')$$

and note that $A \cap (B \cap A') = \varnothing$. This can be easily verified with a Venn diagram. From the property 3 of probability functions we have

$$P(B) = P(A \cup (B \cap A')) = P(A) + P(B \cap A') \geq P(A) \tag{5.17}$$

since $P(B \cap A') > 0$ being a probability. Note that Equation 5.17 is valid since A and $B \cap A'$ are mutually exclusive events. □

Theorem 5.6 $P(A) \leq 1$

Proof Since $A \subset S$ for Theorem 5.5 we have

$$P(A) \leq P(S) = 1$$

Theorem 5.7 *If A and B are events then*

$$P(A \cup B) = P(A) + P(B) - P(A \cap B)$$

Proof This theorem is easy to prove by writing $A \cup B$ as the union of two mutually exclusive events as

$$A \cup B = A \cup (A' \cap B) \tag{5.18}$$

From property 3 of the probability function we have

$$P(A \cup B) = P(A) + P(A' \cap B)$$

However

$$B = (A \cap B) \cup (A' \cap B) \rightarrow P(B) = P(A \cap B) + P(A' \cap B)$$

Now substituting this equation in Equation 5.18 we finally conclude the proof. □

5.7 Conditional Probability

One very important concept in statistics is that of *conditional probability*, that can be defined as follows.

Definition 5.14 The probability of an event A given that the event B has occured is called **conditional probability** and is indicated with

$$P(A|B).$$

We know that in this case A and B must both happen so we can write

$$P(A|B) = \frac{N(A \cap B)}{N(B)} = \frac{N(A \cap B)/N(S)}{N(B)/N(S)} = \frac{P(A \cap B)}{P(B)} \tag{5.19}$$

since in a random experiment we are interested only in those outcomes which are elements of a subset B of the sample space S. Provided of course that $N(B) \neq 0$ (in other words provided that B can happen).

5.8 Bayes Theorem

The Bayes theorem is one of the most important theorems in probability and at the basis of many algorithms in Machine Learning, such as the Naïve Bayes classifier.

Theorem 5.8 (Bayes Theorem) *Given two events A and B, the conditional probability $P(A|B)$ can be written as*

$$P(A|B) = \frac{P(B|A)P(A)}{P(B)} \tag{5.20}$$

Proof The key to the proof is that we can write $P(A \cap B)$ in two ways. It can be written as

$$P(A \cap B) = P(B|A)P(A) \tag{5.21}$$

as we derived in Equation 5.19. But in the same way we can also write it, by symmetry, as

$$P(A \cap B) = P(A|B)P(B) \tag{5.22}$$

By equating the two versions of $P(A \cap B)$ we get the Bayes Theorem. □

5.9 ★ Bayes Error

The **Bayes error** is defined as the lowest possible error rate for any classifier and is sometime called the irreducible error. The best way to understand it, is to discuss the Bayes classifier and show how this classifier reaches the lowest error rate possible for any classifier. In general, suppose we have observations that are described by pair of values: $X \in \mathbb{R}^n$ the observation features and $Y \in \{1, 2, ..., K\}$ the class label. In general the conditional distribution of X given a specific label $Y = r$ will have some form and we can write it generally as

$$P_r(X|y = r) \tag{5.23}$$

In general a classifier is a function $C : \mathbb{R}^n \to \{1, 2, ..., K\}$ that assign to each observation a class label value. The probability of misclassificaiton (sometime called Risk) is then

$$\mathcal{R}(C) = P(C(X) \neq Y) \tag{5.24}$$

The **Bayes Classifier** can then be defined as

$$C_{\text{Bayes}} = \text{argmax}_{r \in \{1,2,\ldots,K\}} P(Y = r|X = x) \tag{5.25}$$

The practical problem, is that it is often impossible to know the probability function P and therefore is impossible to construct the Bayes classifier. We can also define Bayes risk R^* as

$$R^* = P(C_{\text{Bayes}}(X) \neq Y) \tag{5.26}$$

Theorem 5.9 *Given a generic binary classification problem for simplicity, let us define a classifier* h^* *as*

$$h^*(x) = \begin{cases} 1, & \eta(x) \geq 0.5 \\ b, & \eta(x) < 0.5 \end{cases}$$

with $\eta(x) = P(Y = 1|X = x)$. *Then is true that*

1. $R(h^*) = R^*$
2. *For any classifier h, is valid that* $R(h) - R^* = 2\mathbb{E}[|\eta(x) - 0.5|\mathbb{I}_{h(x) \neq h^*(x)}]$
3. $R^* = \mathbb{E}_X[\min(\eta(X), 1 - \eta(X))]$

Proof 1. For a classifier h we can write

$$R(h) = \mathbb{E}_{XY}[\mathbb{I}_{h(x) \neq Y}] = \mathbb{E}_X[\eta(X)\mathbb{I}_{h(x)=0} + (1 - \eta(X))\mathbb{I}_{h(x)=1}]$$

and is easy to see that $R(h)$ is maximised by taking (since it has to be valid $\forall X$)

$$h(x) = \begin{cases} 1, & \eta(X) \geq 1 - \eta(X) \\ 0, & \text{otherwise} \end{cases}$$

that is nothing else than R^*.

2. To prove point 2 note that

$$\begin{aligned} R(h) - R^* &= R(h) - R(h^*) \\ &= \mathbb{E}_X[\eta(X)\mathbb{I}_{h(x)=0} + (1 - \eta(X))\mathbb{I}_{h(x)=1} + \\ &- \eta(X)\mathbb{I}_{h^*(x)=0} - (1 - \eta(X))\mathbb{I}_{h^*(x)=1}] = \\ &= \mathbb{E}_X[|2\eta(X) - 1|\mathbb{I}_{h^*(x) \neq h(x)}] = \\ &= 2\mathbb{E}_X[|\eta(X) - 0.5|\mathbb{I}_{h^*(x) \neq h(x)}] \end{aligned} \tag{5.27}$$

3. That is easy to prove:

$$\begin{aligned} R(h^*) &= \mathbb{E}_X[\eta(X)\mathbb{I}_{h^*(x)=0} + (1 - \eta(X))\mathbb{I}_{h^*(x)=1}] = \\ &= \mathbb{E}_X[\min(\eta(X), 1 - \eta(X))] \end{aligned} \tag{5.28}$$

where the min in the last passage comes from the definition of h^*. □

5.10 ★ Naïve Bayes Classifier

Given an input $\mathbf{x} = (x_1, ..., x_n)$ representing some n features, the Bayes Classifier (BC) assigns to each instance the probabilities

$$P(C_k|x_1, ..., x_n) \tag{5.29}$$

for each of the K possible outcomes of the classes C_k. Using Bayes theorem we can write

$$P(C_k|x_1, ..., x_n) = \frac{P(C_k)P(x_1, ..., x_n|C_k)}{P(x_1, ..., x_n)} \tag{5.30}$$

Definition 5.15 $P(C_k|x_1, ..., x_n)$ is called the **posterior probability**. $P(x_1, ..., x_n|C_k)$ is called the **prior probability**.

Now note that the nominator of Equation (5.30) is simply $P(C_k, x_1, ..., x_n)$ and using the chain rule for the conditional probability we can write

$$\begin{aligned}
P(C_k, x_1, ..., x_n) = {} & p(x_1, ..., x_n, C_k) \\
= & p(x_1|x_2, ..., x_n, C_k)p(x_2, ..., x_n, C_k) \\
= & p(x_1|x_2, ..., x_n, C_k)p(x_2|x_3, ..., x_n, C_k)P(x_3, ..., x_n, C_k) \\
= & ... \\
= & p(x_1|x_2, ..., x_n, C_k)p(x_2|x_3, ..., x_n, C_k)... \\
& ...p(x_{n-1}|x_n, C_k)p(x_n|C_k)p(C_k)
\end{aligned} \tag{5.31}$$

Now we assume (from here the name **Naïve**) that all features x_i are mutually independent. This means that

$$p(x_i|x_{i+1}, ..., x_n, C_k) = p(x_i|C_k) \tag{5.32}$$

therefore we can write

$$P(C_k, x_1, ..., x_n) = p(C_k)\prod_{i=1}^{n} p(x_i|C_k) \tag{5.33}$$

We can also write (with the same thinking)

$$P(x_1, ..., x_n) = P(x_1)P(x_2)...P(x_n) \tag{5.34}$$

Thus Equation 5.30 can be rewritten in the "naïve" assumption as

$$P(C_k|x_1, ..., x_n) = \frac{p(C_k)\prod_{i=1}^{n} p(x_i|C_k)}{P(x_1)P(x_2)...P(x_n)} \tag{5.35}$$

The classifier is then built by choosing as output (or prediction)

$$\hat{y} = \underset{k\in\{1,...,K\}}{\text{argmax}}\ p(C_k)\prod_{i=1}^{n} p(x_i|C_k) \tag{5.36}$$

Note that the denominator is simply ignored since is a constant for each input, therefore does not influence that argmax operator.

Note that the quantities $P(x_i|C_k)$ need in practice to be calculated from the data. For example, let us suppose our dataset is composed of people. If x_i is the age range between 20 and 30 (let's say $x_i = [20, 30]$) we will simply count how many patients in the dataset that have the class C_k (it could be between all sick people for example) are in the age range $[20, 30]$ and then divide by the total number of people in the dataset. In the same way we can easily calculate $P(C_k)$ from the dataset by simply counting how many patients are in class C_k and then by dividing by the total number of patients.

The Naive Bayes classifier, rooted in Bayes' Theorem, is known for its simplicity and effectiveness, particularly in large dataset scenarios. Its main advantages include:

- High efficiency and speed, especially in high-dimensional data scenarios like text classification.
- Serves as an excellent baseline model due to its straightforward implementation.
- The assumption of feature independence might limit its accuracy in real-world scenarios with feature correlation.
- Effective with small sample sizes, despite the potential challenge of high-dimensional feature spaces.
- Suitable for multi-class problems.
- Provides not only classification outcomes but also confidence measures in the form of probability estimates.

Warning **Continuous and Categorical Features**

As usual in machine learning, you do not have to implement the naive bayes classifier by hand, as it is available in the library scikit-learn, for example. But you should pay attention when you have a mix of continuous and categorical features, as in those two cases the classifier must be implemented differently. If you find yourself in such a situation, your best bet is to implement the method from scratch.

In conclusion, Naive Bayes is a versatile and computationally efficient tool in machine learning, ideal for baseline modelling and scenarios where computational resources are limited.

5.11 Distribution Functions

The concept of distributions is fundamental in both statistics and machine learning, as it provides a framework for understanding and modelling the variability and structure of data. In statistics, distributions help to make inferences about population parameters based on sample data, allowing the assessment of probabilities, trends, and patterns. They are essential for hypothesis testing, estimating confidence intervals, and performing regression analysis.

In machine learning, understanding distributions is crucial for feature engineering, model selection, and predictive analysis. Different types of distributions, such as Gaussian or Poisson, guide the choice of algorithms and the interpretation of model outputs. For example, in supervised learning, the nature of the target variable's distribution can influence the selection between regression and classification models. Additionally, distributions play a key role in probabilistic models and Bayesian approaches, where prior distributions are updated with data to obtain posterior distributions. Understanding these concepts allows for more robust and accurate predictions, as well as a deeper comprehension of the underlying data structures and relationships. Let's start with defining the most important concepts that are fundamental in the study of probability distributions.

5.11.1 Cumulative Distribution Function (CDF)

Let us start with the concept of cumulative distribution function (CDF) .

Definition 5.16 (cumulative distribution function) For every random variable X and a probability function $P_X(x)$ we define the *cumulative distribution function* (CDF) denoted by $F_X(x)$ with

$$F_X(x) = P_X(X \leq x) \tag{5.37}$$

That is nothing else than the probability for the random variable of being less than a certain value x. With this definition we can give the following important theorem

Theorem 5.10 ★ *The function $F(x)$ is a CDF if and only if the following three properties are satisfied:*

a. $\lim_{x\to-\infty} F(x) = 0$ *and* $\lim_{x\to\infty} F(x) = 1$
b. $F(x)$ *is a non-decreasing function of* x
c. $F(x)$ *is right-continuous, that is for every number* x_0 *we have* $\lim_{x\downarrow x_0} F(X) = F(x_0)$

Property c is there to make this work even if the x is discrete and not continuous. We will skip here the proof of the theorem. This brings us to the next definition

Definition 5.17 X is continuous if $F_X(x)$ is a continuous function of x. X is discrete if $F_X(x)$ is a step function of x.

And finally let's define what it means for two random variables X and Y to be identically distributed.

Definition 5.18 The random variable X and Y are identically distributed if, for every event A, $P(X \in A) = P(Y \in A)$.

Note that being identically distributed does not mean that X and Y are equal. This statement can also be reformulated in terms of the CDF. In fact, the following theorem is true.

Theorem 5.11 *The following two statements are equivalent:*

a. The random variables X and Y are identically distributed
b. $F_X(x) = F_Y(x)$ for every x.

Proof We will only show that a $\Longrightarrow$ b. Since X and Y are identically distributed, for every x the set $(-\infty, x]$ is in the sigma algebra $\mathcal{B}$ on which the probability function P is defined. Therefore, it is true that

$$F_X(x) = P(X \in (-\infty, x]) = P(Y \in (-\infty, x]) = F_Y(x) \tag{5.38}$$

To prove the inverse b $\Longrightarrow$ a is much more complex and thus we will skip the proof here. □

It is important to learn from that if the CDFs of two random variables are equal, then the two random variables are identically distributed.

5.11.2 Probability Density (PDF) and Mass Functions (PMF)

In this section we will look briefly at the definition of the probability density function (PDF) and of the probability mass function (PMF). Note that the two concepts discussed in this section are similar with the important distinction that the PDF is for continuous variables, while the PMF is for discrete ones.

Definition 5.19 The probability mass function (PMF) of a discrete random variable X is defined by

$$f_X(x) = P(X = x) \tag{5.39}$$

In this case it is easy to see that

$$F_X(x) = P(X \leq x) = \sum_A f_X(x_k) \tag{5.40}$$

assuming that the discrete random variable X can only assume a finite set of N values indicated by x_k with $k = 1, ..., N$, and given $A = \{k | x_k < x\}$.

In the case of a continuous variable the definition is bit trickier. In fact suppose X is a continuous random variable and let's try to calculate $P(X = x)$. To do this note that $\{X = x\} \subset \{x - \epsilon < X \leq x\}$ for any $\epsilon > 0$. Therefore

$$P(X = x) \leq P(X - \epsilon < X \leq x) = F_X(x) - F_X(x - \epsilon) \tag{5.41}$$

for any ϵ. Therefore the following is satisfied

$$0 \leq P(X = x) \leq \lim_{\epsilon \downarrow 0} [F_X(x) - F(x - \epsilon)] = 0 \tag{5.42}$$

So this is not working as expected. To get something usable we need to proceed differently. In the continuous case we need to start by writing $F_X(x)$ as the integral of a mysterious (try to guess what this function will be) function $f_X(t)$

$$F_X(x) = P(X \leq x) = \int_{-\infty}^{x} f_X(t)dt \tag{5.43}$$

Now using the fundamental theorem of calculus, we immediately get

$$\frac{d}{dx}F_X(x) = f_X(x). \tag{5.44}$$

With this in mind we can finally define the PDF.

Definition 5.20 The probability density function (PDF) $f_X(x)$ of a continuous random variable X is the function $f_X(x)$ that satisfy

$$F_X(x) = \int_{-\infty}^{x} f_X(t)dt \tag{5.45}$$

for all x.

The fact that the random variable X has a PDF $f_X(x)$ is indicated with $X \sim f_X(x)$. And now let's give a theorem that will give us the properties that the PMF and PDF must satisfy.

Theorem 5.12 *A function $f_X(x)$ is a PDF (or a PMF) of a random variable X iff*

a. $f_X(x) \geq 0 \;\; \forall x$

b. $\sum_x f_X(x) = 1$ *(PMF) or* $\int_{\mathbb{R}} f_X(x) = 1$ *(PDF)*

Proof **Part 1) If a function $f_X(x)$ is a PDF (or a PMF) then a. and b. are satisfied:**

If $f_X(x)$ is a PDF (or PMF), then from their definition properties a. follows. For b. is easy to note that (here only for continuous random variables)

$$1 = \lim_{x\to\infty} F_X(x) = \int_{\mathbb{R}} f_X(x) \tag{5.46}$$

Part 2) If a. and b. are satisfied then $f_X(x)$ is a PDF (or a PMF):

To prove this we need to prove property a., b. and c. given in Theorem 5.10. Let us start with a. in theorem 5.10. Since $f_X(x)$ is a PDF we have by definition

$$\lim_{x\to-\infty} F_X(x) = \lim_{x\to-\infty} \int_{-\infty}^{x} f_X(t)dt = 0 \tag{5.47}$$

and

$$\lim_{x\to\infty} \leq F_X(x) = \int_{-\infty}^{\infty} f_X(t)dt = 1 \text{ from property b., our assumption.} \tag{5.48}$$

So, property a. in Theorem 5.10 is satisfied. Now, since $f_X(x)$ is always positive (property a., our assumption), then also property b. in Theorem 5.10 (that $F_X(x)$ is a non-decreasing function of x) is satisfied. Finally, since $F_X(x)$ is assumed to be continuous, property c. in theorem 5.10 (that $F_X(x)$ is right-continuous) is automatically satisfied. □

5.12 Expected Values and its Properties

Another very important quantity to define is the expected value of a random variable.

Definition 5.21 The expected value of a random variable X (indicated by $\mathbb{E}(X)$) is its average, and given the PDF (or PMF) it can be evaluated respectively by

$$\mathbb{E}(X) = \int_{\mathbb{R}} x f_X(x) dx \text{ for a continuous variable} \tag{5.49}$$

and

$$\mathbb{E}(X) = \sum_{x} x f_X(x) dx \text{ for a discrete variable} \tag{5.50}$$

where the sum is intended over all possible values of x. If we consider a transformation of a random variable X, as $Y = g(X)$ then the expected value of Ycan be calculated as

$$\mathbb{E}(Y) = \int_{\mathbb{R}} g(x) f_X(x) dx \text{ for a continuous variable} \tag{5.51}$$

and

$$\mathbb{E}(Y) = \sum_{x} g(x) f_X(x) dx \text{ for a discrete variable} \tag{5.52}$$

Theorem 5.13 *Let X and Y be random variables and be a, b, and c constants. Also consider two functions g_1 and g_2 whose expectations exist. Then the following are valid*

1. $\mathbb{E}(aX + bY + c) = a\mathbb{E}(X) + b\mathbb{E}(Y) + c$
2. *If $g_1(x) \geq 0$ for all x, then $\mathbb{E}(g_1(X)) \geq 0$*
3. *If $g_1(x) \geq g_2(x)$ for all x, then $\mathbb{E}(g_1(X)) \geq \mathbb{E}(g_2(X))$*
4. $\mathbb{E}(X + Y) = \mathbb{E}(X) + \mathbb{E}(Y)$

We will skip the proof of this theorem.

5.13 Variance and its Properties

The last important concept that we will examine is *variance*. The variance of a random variable is a measure of the spread of its values.

Definition 5.22 The variance of a random variable X with expected value $\mathbb{E}(X) = \mu$ is indicated with $\text{Var}(X)$ and is defined by the formula

$$\text{Var}(X) = \mathbb{E}[(X - \mu)^2]. \tag{5.53}$$

The variance can be linked to the expectation value of X and X^2. In fact

$$\begin{aligned}\text{Var}(X) &= \mathbb{E}[(X - \mu)^2] \\ &= \mathbb{E}[X^2 - 2X\mathbb{E}[X] + \mathbb{E}[X]^2] \\ &= \mathbb{E}[X^2] - 2\mathbb{E}[X]\mathbb{E}[X] + \mathbb{E}[X]^2 \\ &= \mathbb{E}[X^2] - \mathbb{E}[X]^2.\end{aligned} \tag{5.54}$$

For a continuous random variable X that have a probability density function $f(x)$ the variance is given by the formula

$$\text{Var}(X) = \int_{\mathbb{R}} (x - \mu)^2 f(x) dx \tag{5.55}$$

where

$$\mu = \mathbb{E}(X) = \int_{\mathbb{R}} x f(x) dx \tag{5.56}$$

while for a discrete random variable

$$\text{Var}(X) = \sum_x (x - \mu)^2 f_X(x) \tag{5.57}$$

The variance is linked to the standard deviation σ by the formula

$$\text{Var}(X) = \sigma^2 \tag{5.58}$$

5.13.1 Properties

Here are some of the most important properties of the variance. Some are obvious, some requires proving them.

1. $\text{Var}(X) \geq 0$.
2. $\text{Var}(a) = 0$ with a a constant.
3. $\text{Var}(X + a) = \text{Var}(X)$.
4. $\text{Var}(aX) = a^2\text{Var}(X)$.
5. $\text{Var}(aX + bY) = a^2\text{Var}(X) + b^2\text{Var}(Y) + 2ab\text{Cov}(X, Y)$ with

$$\text{Cov}(X, Y) = \mathbb{E}[(X - \mathbb{E}(X))(Y - \mathbb{E}(Y))] \tag{5.59}$$

this quantity is called the co-variance. Covariance is a statistical measure used to determine the relationship between two variables. It indicates the direction of

the linear relationship between these variables. If the covariance is positive, it means that as one variable increases, the other variable tends to increase as well. If the covariance is negative, it suggests that as one variable increases, the other tends to decrease. A covariance close to zero implies that there is little to no linear relationship between the variables. However, it's important to note that covariance does not indicate the strength of the relationship, nor does it provide any scale, which is why correlation is often preferred for understanding the strength and direction of a linear relationship between two variables.
When the variables satisfy $\text{Cov}(X, Y) = 0$ they are said to be *uncorrelated*. For a set of *uncorrelated* random variables X_i and a set of constants a_i with $i = 1, ..., N$ it is true that

$$\text{Var}\left(\sum_{i=1}^{N} a_i X_i\right) = \sum_{i=1}^{N} a_i^2 \text{Var}(X_i) \tag{5.60}$$

6. if the random variables X_i have the same variance σ^2 then from the previous property follows

$$\text{Var}\left(\frac{1}{N}\sum_{i=1}^{N} X_i\right) = \frac{1}{N^2}\sum_{i=1}^{N} \text{Var}(X_i) = \frac{\sigma^2}{N^2} \tag{5.61}$$

Info Proof of $\text{Var}(aX) = a^2\text{Var}(X)$

As an example of how proofs of the previous properties are done, I will show here how to prove property 4.

Proof Let's start from the definition of variance

$$\text{Var}(X) = \mathbb{E}[(X - \mu)^2] \tag{5.62}$$

Now we need to consider aX. By using the definition, we have

$$\text{Var}(aX) = \mathbb{E}[(aX - a\mu)^2] \tag{5.63}$$

where we have used the property $\mathbb{E}(aX) = a\mathbb{E}(X) = a\mu$ for the linearity of the expectation operator. By using again the linearity property of the expectation, we can easily complete the proof.

$$\begin{aligned} \text{Var}(aX) &= \mathbb{E}[(aX - a\mu)^2] = \mathbb{E}[a^2(X - \mu)^2] = \\ &= a^2\mathbb{E}[(X - \mu)^2] = a^2\text{Var}(X) \end{aligned} \tag{5.64}$$

and that concludes the proof. □

Note that Equation 5.61 is linked to the well-known *central limit theorem*.

5.14 Normal Distribution

We have looked at many general properties and definitions, and to make all this more concrete it is useful to observe how they apply to a real distribution. The best choice is, naturally, the normal distribution, possibly the best known and most used of them all[2]. It is used due to its unique properties and the natural phenomena it describes. Its importance stems from several key aspects.

1. **Ubiquity in Natural Phenomena**: many natural and social phenomena follow a normal distribution making it a powerful tool for modeling and understanding a wide range of real-world data.
2. **Central Limit Theorem**: the Central Limit Theorem states that, under certain conditions, the sum of a large number of random variables, regardless of their distribution, will be approximately normally distributed. This theorem makes the normal distribution a pivotal concept in statistical inference and hypothesis testing.
3. **Simplicity and Mathematical Convenience**: the normal distribution is mathematically tractable, making it easy to calculate probabilities and conduct statistical tests. Its properties allow for the development of powerful statistical methods.
4. **Basis for Other Distributions**: many other important distributions are related to the normal distribution, such as the chi-squared, t, and F distributions. These relationships extend the utility of the normal distribution in statistical modeling and hypothesis testing.
5. **Parameter Estimation**: in statistics and machine learning, the normal distribution is often assumed for the underlying data. This assumption simplifies the estimation of model parameters and enables the use of techniques like maximum likelihood estimation.
6. **Error Modeling**: in many statistical models, especially linear regression, errors are assumed to follow a normal distribution. This assumption facilitates the interpretation of model results and the construction of confidence intervals and hypothesis tests.
7. **Benchmark for Comparison**: the normal distribution serves as a reference point for assessing the distribution of empirical data. Deviations from normality can indicate the presence of skewness, outliers, or other important features in the data.

In this and the next sections we will look at the normal distribution and its properties. It is indicated with the symbol $\mathcal{N}$ and is characterised by two parameters: μ and σ. In this section we will see their meaning. A random variable X that follows a normal distribution will be indicated, as we have done before, with

$$X \sim \mathcal{N}(\mu, \sigma^2) \tag{5.65}$$

we will see why the parameter σ is squared very soon. For the moment ignore this fact. The normal distribution density function, often indicated with $f(x|\mu, \sigma)$, is given by

[2] One distribution to rule them all...

$$f(x|\mu,\sigma) = \frac{1}{\sigma\sqrt{2\pi}} e^{-\frac{1}{2}\left(\frac{x-\mu}{\sigma}\right)^2} \tag{5.66}$$

From Equation (5.66) it should be immediately clear that it is symmetric in $x - \mu$, or in other words is centered at $x = \mu$. It goes to zero for $x \to \pm\infty$ and has a bell shape. You can see it for $\mu = 0$ and $\sigma = 1$ in Figure 5.1. Note that $\mathcal{N}(0, 1)$ is called the **standard normal distribution**.

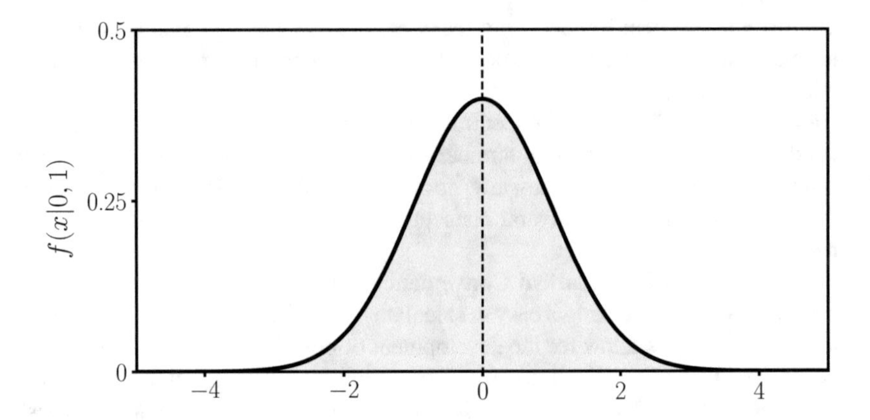

Fig. 5.1 The normal distribution density function for $\mu = 0$ and $\sigma = 1$. The normal distribuion with those parameters is called a *standard* normal distribution.

Now, let us explore the significance of the parameters μ and σ. Let us start by calculating the average of a random variable that follows the normal distribution. We need to calculate

$$\begin{aligned}
\mathbb{E}(X) &= \int_{\mathbb{R}} x f(x|\mu,\sigma^2) = \int_{\mathbb{R}} x \frac{1}{\sigma\sqrt{2\pi}} e^{-\frac{1}{2}\left(\frac{x-\mu}{\sigma}\right)^2} dx = \\
&= \frac{1}{\sigma\sqrt{2\pi}} \int_{\mathbb{R}} x e^{-\frac{1}{2}\left(\frac{x-\mu}{\sigma}\right)^2} dx = \left\{ s = \frac{x-\mu}{\sigma} \Rightarrow \sigma ds = dx \right\} = \\
&= \frac{1}{\sqrt{2\pi}} \int_{\mathbb{R}} (\sigma s + \mu) e^{-\frac{1}{2}s^2} ds = \underbrace{\frac{\sigma}{\sqrt{2\pi}} \int_{\mathbb{R}} s e^{-\frac{1}{2}s^2} ds}_{A} + \\
&+ \underbrace{\frac{\mu}{\sqrt{2\pi}} \int_{\mathbb{R}} e^{-\frac{1}{2}s^2} ds}_{B}
\end{aligned} \tag{5.67}$$

Now note that the integral indicated with A is zero, given the symmetry of the function under the integral sign. Also note that we can use the known result.

$$\int_{\mathbb{R}} e^{-\frac{1}{2}s^2} = \sqrt{2\pi} \tag{5.68}$$

to get the final result

$$\mathbb{E}(X) = \mu \tag{5.69}$$

The parameter μ is the expected value of a random variable that follows a normal distribution.

Info Proof of Equation (5.68)

Let us define

$$A = \int_{\mathbb{R}} e^{-\frac{1}{2}s^2} \tag{5.70}$$

The trick to get the result in Equation 5.68 is to calculate A^2 and not A. In fact we can write

$$A^2 = \int_{\mathbb{R}} dx \int_{\mathbb{R}} dy e^{-\frac{1}{2}x^2} e^{-\frac{1}{2}y^2} = \int_{\mathbb{R}} dx \int_{\mathbb{R}} dy e^{-\frac{1}{2}(x^2+y^2)} \tag{5.71}$$

Now we move to polar coordinates with the change of variables

$$\begin{cases} x &= r\cos(\theta) \\ y &= r\sin(\theta) \end{cases} \tag{5.72}$$

with r going from 0 to ∞ and θ going from 0 to 2π. Since we are making a change of variables (more than one), we need to calculate the Jacobian J. The reader should know that the following formula is valid

$$dxdy = |J| dr d\theta \tag{5.73}$$

where $|J|$ indicates the determinant of the Jacobian. In this case it is easy to see that

$$J = \begin{pmatrix} \frac{\partial x}{\partial r} & \frac{\partial x}{\partial \theta} \\ \frac{\partial y}{\partial r} & \frac{\partial y}{\partial \theta} \end{pmatrix} = \begin{pmatrix} \cos(\theta) & -r\sin(\theta) \\ \sin(\theta) & r\cos(\theta) \end{pmatrix} \tag{5.74}$$

therefore

$$|J| = r\cos^2(\theta) + r\sin^2(\theta) = r \tag{5.75}$$

With this result we can re-write A^2 as

$$A^2 = \int_0^{\infty} dr \int_0^{2\pi} d\theta r e^{-\frac{1}{2}r^2} = 2\pi \int_0^{\infty} r e^{-\frac{1}{2}r^2} dr \tag{5.76}$$

that with the change of variable $s = r^2/2 \rightarrow ds = rdr$ can be easily calculated

$$A^2 = 2\pi \int_0^{\infty} e^{-s} ds = 2\pi \Rightarrow A = \sqrt{2\pi} \tag{5.77}$$

The reader should be able to do the integral in Equation (5.77) easily. This concludes the proof.

To understand the parameter σ we need to calculate the variance of a random variable following a normal distribution. The integral to calculate is

$$\text{Var}(X) = \int_{\mathbb{R}} (x-\mu)^2 f(x|\mu,\sigma^2) = \int_{\mathbb{R}} (x-\mu)^2 \frac{1}{\sigma\sqrt{2\pi}} e^{-\frac{1}{2}\left(\frac{x-\mu}{\sigma}\right)^2} dx \tag{5.78}$$

since we now know that $\mathbb{E}(X) = \mu$. Let us evaluate the integral now with the same change of variable we have done before, namely $s = (x-\mu)/\sigma$.

$$\text{Var}(X) = \int_{\mathbb{R}} s^2\sigma^2 \frac{1}{\sigma\sqrt{2\pi}} e^{-\frac{1}{2}s^2} \sigma ds = \sigma^2 \underbrace{\frac{1}{\sqrt{2\pi}} \int_{\mathbb{R}} s^2 e^{-\frac{1}{2}s^2} ds}_{A} \tag{5.79}$$

we can show that $A = 1$. This tells us that the parameter σ^2 is nothing else than the variance of a random variable that follows a normal distribution:

$$\text{Var}(X) = \sigma^2 \tag{5.80}$$

Info Proof that $A = 1$ in Equation (5.79)

We want to prove that the following equation is valid

$$A = \frac{1}{\sqrt{2\pi}} \int_{\mathbb{R}} x^2 e^{-\frac{1}{2}x^2} dx = 1. \tag{5.81}$$

To do this we can use a neat trick. Let's start with the integral

$$I(\alpha) = \int_{\mathbb{R}} e^{-\alpha x^2} dx \tag{5.82}$$

and note that, similar to what we have done in the Proof of Equation ((5.68)), it is easy to show that

$$I(\alpha) = \int_{\mathbb{R}} e^{-\alpha x^2} dx = \sqrt{\frac{\pi}{\alpha}}. \tag{5.83}$$

Now we can take the derivative[a] of $I(\alpha)$

$$\begin{aligned} \frac{dI(\alpha)}{d\alpha} &= \frac{d}{d\alpha} \int_{\mathbb{R}} e^{-\alpha x^2} dx = \int_{\mathbb{R}} \frac{d}{d\alpha} e^{-\alpha x^2} dx = \\ &= -\int_{\mathbb{R}} x^2 e^{-\alpha x^2} dx \end{aligned} \tag{5.84}$$

But from Equation (5.83) we know that

$$\frac{dI(\alpha)}{d\alpha} = -\frac{\sqrt{\pi}}{2}\alpha^{-3/2} \tag{5.85}$$

by equating Equations (5.84) and (5.85) we get

$$\int_{\mathbb{R}} x^2 e^{-\alpha x^2} dx = \frac{\sqrt{\pi}}{2}\alpha^{-3/2} \tag{5.86}$$

and by choosing $\alpha = 1/2$ we get the final result

$$\int_{\mathbb{R}} x^2 e^{-x^2/2} dx = \sqrt{2\pi} \tag{5.87}$$

This concludes the proof.

---

[a] We will not discuss here the applicability of exchanging the derivative and the integral sign. This kind of discussion goes beyond the scope of the book.

Now let's calculate the CDF of a normal distribution. Let us start with the CDF of a standard normal distribution (remember is a normal distribution with $\mu = 0$ and $\sigma = 1$). This is usually indicated with $\Phi(x)$ and is by definition

$$\Phi(x) = \int_{-\infty}^{x} f(x|0,1) = \frac{1}{\sqrt{2\pi}}\int_{-\infty}^{x} e^{-t^2/2} dt \tag{5.88}$$

Note that we cannot express this integral in closed form. So in case you need to evaluate it, you need to do it numerically. $\Phi(x)$ is simply the area under $f(x|0,1)$ from $-\infty$ to x, that you can see in Figure 5.2 grayed out. The CDF of a general normal distribution can be expressed with $\Phi(x)$ as

$$\begin{aligned} F(x) &= \frac{1}{\sigma\sqrt{2\pi}}\int_{-\infty}^{x} e^{-\left(\frac{t-\mu}{\sigma}\right)^2/2} dt = \left\{s = \frac{t-\mu}{\sigma} \Rightarrow \sigma ds = dt\right\} \\ &= \frac{1}{\sqrt{2\pi}}\int_{-\infty}^{\frac{x-\mu}{\sigma}} e^{-s^2/2} ds = \Phi\left(\frac{x-\mu}{\sigma}\right) \end{aligned} \tag{5.89}$$

In Table 5.1 there is a summary of the results so far. The same results can be obtained for various other distributions.

5.15 ■ Other Distributions

The number of distributions that have been studied is very large. Here is a list, as a reference, of some of the many available (for completeness, I included also the

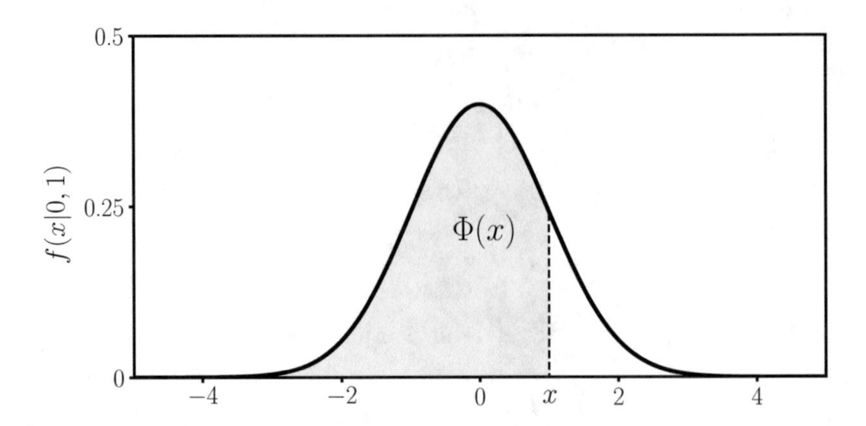

Fig. 5.2 $\Phi(x)$ is the value of the area grayed out. Note that $\Phi(x)$ is the symbol for the CDF for a standard normal distribution not for a generic normal distribution.

Table 5.1 Relevant formulas related to the normal distribution.

Function	Formula
$f(x\|0,1)$	$\frac{1}{\sqrt{2\pi}}e^{-\frac{1}{2}x^2}$
$f(x\|\mu,\sigma)$	$\frac{1}{\sigma\sqrt{2\pi}}e^{-\frac{1}{2}\left(\frac{x-\mu}{\sigma}\right)^2}$
$\Phi(x)$ (CDF Standard Normal Distribution)	$\Phi(x) = \frac{1}{\sqrt{2\pi}}\int_{-\infty}^{x} e^{-t^2/2}dt$
$F(x)$ (CDF Normal Distribution)	$F(x) = \frac{1}{\sigma\sqrt{2\pi}}\int_{-\infty}^{x} e^{-\left(\frac{t-\mu}{\sigma}\right)^2/2}dt = \Phi\left(\frac{x-\mu}{\sigma}\right)$

Gaussian distribution that we just discussed). The list includes the probability or mass density function for each of the distributions.

1. **Normal Distribution (Gaussian Distribution)**: Used for modelling continuous data where observations cluster around a mean. Commonly applied in natural and social sciences to represent real-valued random variables whose distributions are not known.

   Probability Density Function

$$f(x|\mu,\sigma) = \frac{1}{\sigma\sqrt{2\pi}}e^{-\frac{1}{2}\left(\frac{x-\mu}{\sigma}\right)^2}$$

   where μ is the mean and σ is the standard deviation.

2. **Binomial Distribution**: Applied when modeling the number of successes in a fixed number of independent Bernoulli trials (e.g., coin tosses), each with the same probability of success.
   Probability Mass Function

$$P(X = k) = \binom{n}{k} p^k (1 - p)^{n-k}$$

   where n is the number of trials, k is the number of successes, and p is the probability of success on a single trial.
3. **Poisson Distribution**: Used for modeling the number of times an event occurs in a fixed interval of time or space. Suitable for events that occur independently and at a constant average rate.
   Probability Mass Function

$$P(X = k) = \frac{\lambda^k e^{-\lambda}}{k!}$$

   where λ is the average number of events in the interval.
4. **Exponential Distribution**: Commonly used to model the time until the next event in a Poisson process, such as the time until a radioactive particle decays or the time between arrivals in a queue.
   Probability Density Function

$$f(x|\lambda) = \lambda e^{-\lambda x}$$

   for $x \geq 0$, where λ is the rate parameter.
5. **Uniform Distribution**: Describes an experiment where all outcomes are equally likely. Used when there is no bias or preference for any outcome over another.
   Probability Density Function

$$f(x|a, b) = \frac{1}{b - a}$$

   for $a \leq x \leq b$, where a and b are the lower and upper bounds of the distribution, respectively.
6. **Chi-Squared Distribution**: Frequently used in hypothesis testing and confidence interval estimation, particularly for variance. It is the distribution of a sum of the squares of k independent standard normal random variables.
   Probability Density Function

$$f(x|k) = \frac{1}{2^{k/2}\Gamma(k/2)} x^{k/2-1} e^{-x/2}$$

   for $x > 0$, where k is the degrees of freedom and $\Gamma(x)$ is the gamma function. In the case of an integer $\Gamma(n) = (n - 1)!$.

7. **t-Distribution**: Useful for estimating population parameters for small sample sizes or unknown variances. Commonly used in hypothesis testing, especially for the mean of a normally distributed population when the sample size is small.
   Probability Density Function

$$f(x|v) = \frac{\Gamma((v+1)/2)}{\sqrt{v\pi}\,\Gamma(v/2)} \left(1 + \frac{x^2}{v}\right)^{-(v+1)/2}$$

   where v is the degrees of freedom.
8. **Beta Distribution**: Used to model variables that are constrained to intervals, such as proportions and percentages. It is useful in Bayesian analysis and for modeling success probabilities in binomial experiments.
   Probability Density Function

$$f(x|\alpha, \beta) = \frac{x^{\alpha-1}(1-x)^{\beta-1}}{B(\alpha, \beta)}$$

   for $0 \leq x \leq 1$, where α and β are shape parameters and $B(\alpha, \beta)$ is the Beta function.
9. **Gamma Distribution**: Often used to model waiting times, like the time until death in a biological organism or system failure in engineering. It generalizes the exponential distribution for non-negative values.
   Probability Density Function

$$f(x|\alpha, \beta) = \frac{\beta^\alpha}{\Gamma(\alpha)} x^{\alpha-1} e^{-\beta x}$$

   for $x \geq 0$, where α is the shape parameter and β is the rate parameter.

5.16 The MSE and its Distribution

So far, we have discussed very fundamental concepts in statistics. Let us explore how advanced statistics plays an important role in machine learning using a simple example. We will consider, as usual, a dataset composed of tuples $\{x_i, y_i\}_{i=1}^N$ with $x_i, y_i \in \mathbb{R}$. The data will be generated with the formula $y_i = x_i^2 + \epsilon_i$ where ϵ_i is a random number between -3.5 and 3.5. We will select randomly 100 points from the dataset and train a model. The results shown have been obtained with a gradient boosting regressor[3]. Note the model I have chosen is not relevant for this discussion. Since we can generate as many points as we want, let us generate 3000 test datasets of various sizes. For each of the 3000 test datasets we calculate the MSE and thus we will end up with 3000 values for the MSE. Normally you have only **one** test data sets, but please bear with me, as I would like to show you something. At the end of the

[3] The model have been trained with 500 estimators, maximum depth of 4, minimum sample split of 5, a learning rate of 0.01 and as loss the squared error.

chapter, we will discuss this point again. Let us consider the following test dataset sizes: 10, 50, 75, 100, 500 and 1000. For each of the size we will generate 3000 datasets and evaluate the MSE on them. Can we say something about the distribution of the 3000 values of the MSE? How is influenced by the dataset size? For practical reasons we will indicate the size of the test dataset with s_t. In Figure 5.3 you can see the distribution of the MSE values for $s_t = 10$. The distribution is asymmetric and

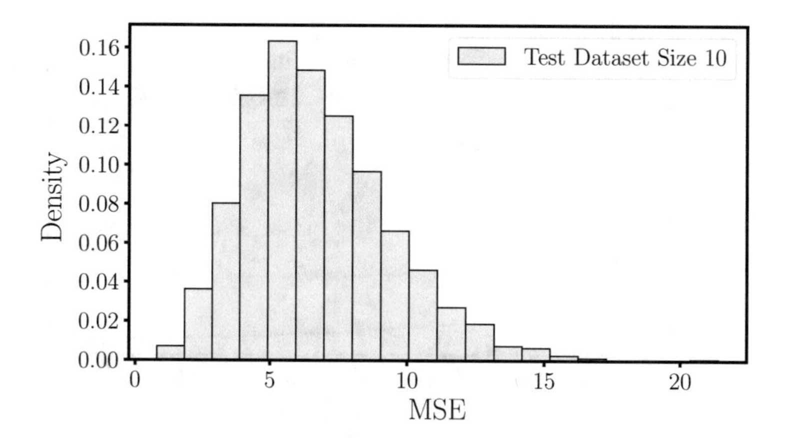

Fig. 5.3 The distribution of MSE values evaluated over 3000 test datasets each of size $s_t = 10$.

have a long tail in the direction of higher values. We have no reason to think that this distribution should be symmetrical. In Figure 5.4 we see the same distribution for $s_t = 1000$. Suddenly, the distribution is symmetric. But what exactly is going on? It

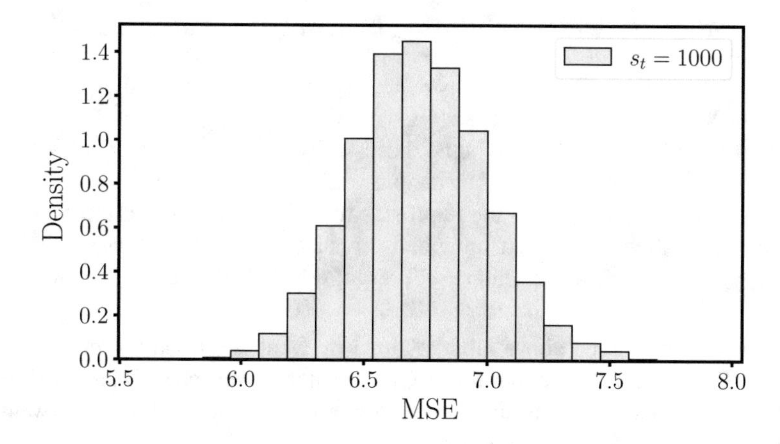

Fig. 5.4 The distribution of MSE values evaluated over 3000 test datasets each of size $s_t = 1000$.

turns out that the distributions for larger (what *larger* means is still not discussed) s_t are not only symmetrical, but gaussian ones. Now, check Figure 5.5. In it you will see the standard deviation of the MSE on the y-axis, and the test dataset size s_t on the x-axis. The dashed line is a theoretical prediction. If you are wondering, yes, it is actually possible to predict how the standard deviation of the MSE varies while varying s_t. All this is possible thanks to statistics. In the next section, we will do

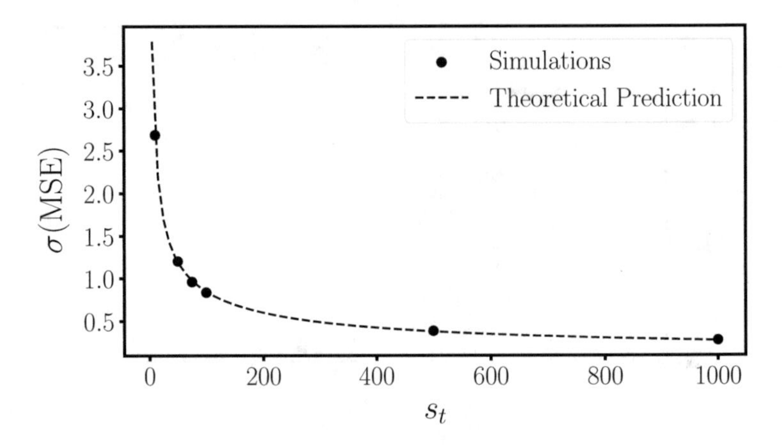

Fig. 5.5 The points are the standard deviation of the MSE ploted versus the test dataset size s_t. The dashed line is a theoretical prediction, as explained in the text.

that and discuss the necessary (more) advanced statistical concepts to be able to get the dashed line in Figure 5.5. This result will be widely discussed in Chapter 7 on model validation.

Info Statistics when Validating Machine Learning Models

Since giving the MSE for various dataset splits or in what is called k-Fold Cross Validation (more on that in Chapter 7) is a standard practice, this result is quite important. You do not necessarily have to remember **why** the MSE distributions behave as they do, but you should remember what kind of behaviour they have, to be able to provide the correct statistical information when validating models. At this point this discussion may still seem a bit foggy, but don't worry. In Chapter 7 everything will be discussed in detail.

The results discussed above are explained by what is termed the central limit theorem (CLT). The next sections on the CLT are more advanced and can be skipped. Finally, we summarise the results without the proofs for those who do not want to go through all the mathematical steps.

5.16.1 ★ Moment Generating Functions

To be able to prove the central limit theorem in the next section, we need the concept of moment generating functions (MGFs). MGFs are a fundamental concept in the field of probability theory and statistics. They provide a powerful tool for characterising probability distributions and have applications in various statistical analyses and theoretical derivations. The moment-generating function of a random variable offers a unique way to encode all the moments (i.e., expected values of powers) of the variable's probability distribution.

The moment-generating function of a random variable X is defined as follows:

$$M_X(t) = \mathbb{E}[e^{tX}] = \int_{-\infty}^{\infty} e^{tx} f_X(x)\, dx \quad \text{(for continuous variables)} \tag{5.90}$$

$$M_X(t) = \mathbb{E}[e^{tX}] = \sum_x e^{tx} P(X = x) \quad \text{(for discrete variables)} \tag{5.91}$$

where $\mathbb{E}$ denotes the expected value, t is a real number, $f_X(x)$ is the probability density function of X for continuous variables, and $P(X = x)$ is the probability mass function for discrete variables. The MGF has several important properties and applications:

- **Existence**: The MGF exists if it is finite for some interval around $t = 0$. Its existence ensures that all moments of the random variable exist.
- **Uniqueness**: If the MGF of a random variable exists, it uniquely determines the probability distribution of the variable.
- **Moments**: The n-th moment of X can be obtained by differentiating $M_X(t)$ n times and evaluating the result at $t = 0$:

$$\mathbb{E}[X^n] = M_X^{(n)}(0) = \frac{d^n}{dt^n} M_X(t)\bigg|_{t=0} \tag{5.92}$$

- **Summation of Independent Variables**: If X and Y are independent random variables, then the MGF of their sum $X + Y$ is the product of their individual MGFs.

$$M_{X+Y}(t) = M_X(t) \cdot M_Y(t) \tag{5.93}$$

- **Convergence**: MGFs are used to prove convergence in the distribution. If the MGFs of a sequence of random variables converge to the MGF of a random variable, the sequence converges in distribution to that variable.

Moment-generating functions are used extensively in statistical analyses:

- **Deriving Distributions**: They are instrumental in deriving the distributions of functions of random variables, especially sums and linear combinations.
- **Characterization**: MGFs can be used to characterise some well-known distributions, such as normal, exponential, and Poisson distributions.

- **Hypothesis Testing**: In hypothesis testing, MGFs aid in deriving the sampling distribution of test statistics under the null hypothesis.
- **Parameter Estimation**: In parametric statistics, MGFs can help in estimating parameters of a distribution.

Understanding moment-generating functions is crucial for anyone studying probability theory and statistical analysis, as they provide a comprehensive framework for dealing with random variables and their distributions.

5.16.2 ★ Central Limit Theorem

One of the key theorems in statistics is the Central Limit Theorem (CLT). Its first version dates back to 1811 and several versions with various hypothesis are available. In this book we will look at one variation and at one proof that is easy enough to understand with the material presented in this chapter. Let us start with the formulation of a version of the CLT with the strictest hypothesis for its validity.

Theorem 5.14 (Central Limit Theorem) *Consider X_1, X_2, ..., X_n n random variables independent and identically distributed with mean 0 and standard deviation σ^2. Let us also suppose that the moment generating function $M(t)$ is defined in a neighbourhood of zero. Let us define*

$$S_n = \sum_{i=0}^{n} X_i \tag{5.94}$$

Then

$$\lim_{n\to\infty} P\left(\frac{S_n}{\sigma\sqrt{n}} \le x\right) = \Phi(x) \tag{5.95}$$

where $\Phi(x)$ is the CDF of the standard normal distribution. Formulated differently, this theorem tells us that $S_n/(\sigma\sqrt{n})$ in the limit of $n \to \infty$ will tend to follow a standard normal distribution.

Proof The proof is not complicated but it is a bit tedious. All passages are explained in detail to make it as clear as possible and at the same time rigorous. Let us start. We define

$$Z_n = \frac{S_n}{\sigma\sqrt{n}}. \tag{5.96}$$

Since S_n is the sum of random variables that are independent (see previous section) then

$$M_{S_n}(t) = [M(t)]^n \tag{5.97}$$

and of course

$$M_{Z_n}(t) = \left[M\left(\frac{t}{\sigma\sqrt{n}}\right)\right]^n \tag{5.98}$$

Since $M(t)$ is defined we can write its Taylor's expansion around 0

$$M(s) = M(0) + sM'(0) + \frac{1}{2}s^2 M''(0) + \epsilon_s \tag{5.99}$$

where the following must be satisfied (Taylor's theorem)

$$\lim_{s\to 0} \frac{\epsilon_s}{s^2} = 0 \tag{5.100}$$

Now from our hypothesis (zero mean and standard deviation equal to σ) we have $M'(0) = 0$ and $M''(0) = \sigma^2$. Additionally

$$M(0) = \int_{\mathbb{R}} f_X(x)dx = 1 \tag{5.101}$$

where $f_X(x)$ is the PDF of the random variables X_i and therefore has an integral equal to one. Now we can rewrite Equation (5.98) by using Equation (5.99).

$$M_{Z_n}(t) = \left[1 + \frac{t^2}{2n} + \epsilon_n\right]^n \tag{5.102}$$

To make the proof easier to follow, notice that the following is true

$$\lim_{n\to\infty} M_{Z_n}(t) = \lim_{n\to\infty} \left[1 + \frac{t^2}{2n} + \epsilon_n\right]^n = \lim_{n\to\infty} \left[1 + \frac{t^2}{2n}\right]^n \tag{5.103}$$

This is not obvious *at all* and needs to be proved. We will skip the proof in this book. We will use the very well-known result that if $b_n \to b$ for $n \to \infty$ then

$$\lim_{n\to\infty} \left(1 + \frac{b_n}{n}\right)^n = e^b \tag{5.104}$$

by using this result and Equation (5.103) we finally get

$$\lim_{n\to 0} M_{Z_n}(t) = e^{t^2/2} \tag{5.105}$$

This is the moment generating function of a normal distribution with mean zero and standard deviation of 1 (I hope you will believe me on this). Thus, we have proven Equation (5.95).

Warning Some Additional Details about the Proof

In the previous proof, we have used, without explicitly saying it, an important result about MGFs. In fact, even if two random variables have the same MGFs, it does not mean that they are identically distributed. So some care is required. An important theorem (that we will not prove here) states the following. Consider two random variables X and Y that have moment generating functions $M_X(t)$ and $M_Y(t)$. Suppose that all moments exists. If $M_X(t) = M_Y(t)$ for all t in some neighbourhood of 0, then $F_X(t) = F_Y(t)$ and therefore the two variables are identically distributed. This is required in the last passage of the proof. Only with this result can we say that since the limit of $M_{Z_n}(t)$ for $n \to \infty$ is the MGF of the standard normal distribution. This is the main reasons why we need to say something about MGFs in the theorem hypothesis. If you are interested in a more detailed discussion about this point you can check Section 2.3 of [1]. Note that the CLT can be proved without using MGFs and thus it is also valid if the MGFs are not always defined. This can be achieved with characteristic functions but the proof, although similar, is more delicate and require more attention. We will not give it here. Just for the reader's reference, the characteristic function of a random variable is a complex valued function defined by

$$\varphi(t) = \mathbb{E}[e^{itX}] \tag{5.106}$$

where of course $i = \sqrt{-1}$.

5.17 Central Limit Theorem without Mathematics

The Central Limit Theorem (CLT) is one of the most powerful and fundamental concepts in statistics and probability theory. It provides a bridge between probability distributions and enables us to understand why many processes in nature and human-made systems tend to follow a normal distribution, even if the underlying processes do not.

The CLT states that when independent random variables are added, their properly normalized sum tends toward a normal distribution (informally a "bell curve"), even if the original variables themselves are not normally distributed. The key elements of this theorem are independence and summing up of random variables. Imagine you're taking a large number of random samples from any population, say the heights of people, the weight of apples, or even the amount of time people spend on a website. Each of these samples can have any distribution with its own mean and variance. Now, if you calculate the average of each sample and plot these averages, the surprising result given by the CLT is that this new distribution of averages will resemble a normal distribution.

Each sample's average is a combination of many random influences. When these influences combine, their individual peculiarities (like skewness or being lopsided) tend to cancel out. What's left is the collective tendency, which, as the CLT shows, is a normal distribution. The theorem becomes more accurate as the size of the samples increases. The magic number often cited in statistics is 30. With a sample size above 30, the CLT starts to hold true, and the distribution of sample means becomes increasingly normal.

This theorem is the reason why normal distribution is so prevalent in statistical methods. It allows us to make inferences about population parameters even when the population itself doesn't strictly follow a normal distribution. For example, it underpins many statistical tools, such as confidence intervals and hypothesis tests. In conclusion, the Central Limit Theorem is a cornerstone of probability theory and statistics. It's the reason why the normal distribution is so fundamental in the field. It shows that in many situations, natural or experimental, the sum or average of a large number of random variables will tend to be normally distributed, regardless of the distribution from which the variables are drawn.

References

1. George Casella and Roger L Berger. *Statistical inference*. Cengage Learning, 2021.

Chapter 6
Sampling Theory (a.k.a. Creating a Dataset Properly)

Surveys show that surveys never lie.

Natalie Angier

This chapter dives into sampling theory, which is very important if you are working on anything from digital technology to health sciences. Think of it as choosing the best pie slices that give you a real taste of the whole thing. This helps you make solid guesses and build or test machine learning models that actually work according to what you need to find out. Sampling theory deals with the challenge of creating representative subsets of a larger population. First, we discuss why it is critical to have clear research questions and hypotheses before you start collecting data. Getting the right ones helps figure out exactly which data you need to prove or disprove your point. Then, we break down survey sampling into two types: non-probability and probability sampling. Each type has its own way of picking data samples and is used for different reasons depending on what you are trying to study. We also discuss how to group your data (stratification and clustering) and look at different ways to pick random samples, either by making sure that each piece of data gets chosen only once or by possibly choosing the same piece more than once (with and without replacement).

6.1 Introduction

Sampling theory has applications in various fields. For example, in digital signal processing, it deals with the problem of reconstructing and converting signals. The theory is particularly useful for understanding how continuous-time signals (analogue signals) can be converted into discrete-time signals (digital signals) for processing by digital systems and then reconstructed back into continuous forms if necessary. It is used in control systems where, by smartly sampling sensors, it is possible to develop algorithms to make efficient and timely decisions. Sampling theory is also used in medical imaging, telecommunications, finance, and many other fields. Most importantly, for us, in statistics, sampling theory guides the selection of representative subsets from larger populations for analysis, ensuring that statistical

U. Michelucci, *Fundamental Mathematical Concepts for Machine Learning in Science*,
https://doi.org/10.1007/978-3-031-56431-4_6

inferences about the population are valid and reliable. In this chapter, we discuss this last application.

Typically, when studying machine learning, you will find yourself using datasets that have been previously prepared and curated by others. However, when starting a new research project, you may be faced with the task of creating a completely new dataset that will be used to answer a specific research question. Acquiring data is trickier than you can imagine, and knowing the fundamentals of sampling will help you assess bias, representativeness, and much more about your data. It will crucially help you in deciding **what** kind, **how much**, **how many types** of samples to acquire for a given research question.

Let us consider a couple of examples on how acquiring data may not be as easy as you may think.

- A researcher wants to apply computer vision to create a tumour classification algorithm on MRI scans. Possible questions regarding data acquisition may be: do we need to have all age ranges between patients? What kind of tumour do we want to classify? Do we want to exclude any comorbidity in patients? How many scans do we need for each patient? Do we want to have as many scans with and without tumours? The number of questions goes on, and having some strategy on how to ask the right questions in the first place is a fundamental skill for any machine learning researcher.
- A company wants to study chemical substance fingerprints obtained by optical means (for example, Raman scattering or fluorescence in case you know the methods). What kinds of substances should we consider? What measurement techniques? What parameters should be used for the instruments? The list of questions goes on.

These two simple examples should give you at least an intuitive idea of how many different questions you must answer when defining what new data to acquire for a machine learning project. In this chapter, we review the fundamental concepts of sampling that will help you design a proper data acquisition strategy for your projects.

6.2 Research Questions and Hypotheses

We now discuss7 two important concepts that drive the creation of datasets: research questions (RQ) and hypotheses. Research projects often start with a dataset. Someone finds or is given a dataset and is asked to predict something. Sounds familiar? Maybe the professor you are working with got a dataset from some other research group and would like you to *try something with machine learning*, or directly ask you to train some algorithm to predict something.

This is how most of these projects start. If you find yourself in this situation, take a step back and think about the problem you are trying to solve. What questions are you trying to answer? What hypotheses do you have? It is important to clarify some terminology and explain what a research question or hypothesis is, why you need one, and how designing a good hypothesis to verify (or disprove) is mandatory for good dataset creation. Before getting to hypotheses, let us start by asking what a research question is.

6.2.1 Research Questions

A good research project starts **always** with one (or multiple) research question (RQ). We can loosely define it as follows.

Definition 6.1 (Research Question) A research question is a concise, focused inquiry formulated to address a specific concern or knowledge gap within a broader topic area.

Since this definition is quite generic, let us give three examples to aid in your understanding of the concept.

- How does the introduction of non-native plant species affect biodiversity in urban green spaces?
- Can we predict the onset of diabetes from the medical history of patients?
- What impact does the integration of technology in the classroom have on student engagement and learning outcomes?

By looking at the examples, you will realise that the questions are general in nature and cannot really be answered with some precise number (take the last question, what does *impact* mean, and how can it be measured?). They guide the research process, determining the direction of the study, and can give some hints on how to decide what types of data or methods need to be collected and used. In any research project, you should start with a research question (or multiple ones, if is relevant). This is the typical approach that you use if you are working on your thesis (Bachelor, Master, or even a PhD) by the way. A good RQ gives the context of the research project, hints at its impact, and highlights its importance. It should be understood by non-experts and inspire. Additionally, it typically allows for a wide range of outcomes.

As its name implies, it is good practice to formulate it as a **question**. Avoid statements that are not formulated as questions.

6.2.2 Hypothesis

Once you formulate your RQ, you will need hypotheses. Something you can disprove or verify[1] Hypotheses can be loosely defined as the following.

[1] Karl Popper, the famous philosopher would disagree on the verification of a scientific hypothesis, but we will skip this discussion here.

Definition 6.2 (Definition) A hypothesis is a prediction about the relationship between two or more variables. It can be described as an educated guess about what happens in an experiment. Researchers tend to use hypotheses when significant knowledge on the subject is already available. The hypothesis is based on existing knowledge. After the hypothesis is developed, the researcher can develop or gather data, analyse them, and use them to support or negate the hypothesis.

Some examples are as follows.

- Global warming has increased the sea level by 1 cm in the past 10 years on average globally.
- The number of cars on the road each day on average in Zürich has decreased by 5% after the COVID year.

You should immediately see the difference between a hypothesis and a research question. While a RQ is written as a question (hence the name), a hypothesis is always written as a statement that can be verified or disproved. An hypothesis is the fundamental building block that allows you to design experiments to test the hypothesis itself. In other words, it means that getting the right data is a consequence of a well-thought hypothesis.

To summarise what we discussed when starting a new research project, you should proceed according to the following steps.

1. Formulate one (or multiple, but not more than 2-3) research question.
2. Formulate a series of hypotheses that will help you answer your research questions.
3. Design experiments to verify or disprove the hypotheses formulated.

Info Experiments in Machine Learning

An **experiment** is a methodical procedure carried out with the objective of verifying or falsifying one or multiple hypotheses. Experiments involve manipulating one or more variables to determine their effect on a certain outcome. In machine learning, an experiment consists of building a complete pipeline that starts with data processing and finishes with model validation. For example, your hypothesis might be that *with a specific set of MRI images (we will not discuss too many details here) you can predict a specific medical condition with an accuracy higher than a certain threshold with machine learning approaches*. The experiments you can design to verify or disprove this hypothesis could consist of building a set of pipelines that train a set of machine learning models and test their performance by varying things like model type, data normalisation, etc.
An **experiment** in machine learning can be loosely (and surely not precisely) defined as building a complete pipeline that starts with data processing and ends with model validation.

During the work of points 2 and 3 you will find yourself having enough information to be able to design your data collection strategy. Let us discuss further how to do that.

6.2.3 Relevance of Hypothesis and Research Questions in Machine Learning

We have now clarified what a research question (RQ) and a hypothesis entail, but why do they matter? Primarily, these elements act as essential tools to organise your research project, regardless of whether it involves machine learning. Moreover, focusing specifically on sampling theory, only through a well-written hypothesis you can create a dataset that effectively tests your hypothesis. The data you employ, choose, or manipulate are directly shaped by your hypothesis.

Let us simplify this with an example. Imagine that your research question (RQ) is, "What is the effect of weight on cardiovascular diseases?" A plausible hypothesis could be "Men with a BMI greater than 25 are more likely to have cardiovascular disease". With this hypothesis, your task is to collect weight and height data from as many men as possible, both above and below a BMI of 25, and to note if they have a cardiovascular disease (CVD). It is also important to aim for a balanced dataset, including a similar number of men with and without CVD, as discussed in Chapter 8 on handling unbalanced data.

When planning your data collection, you might consider focusing on a specific age group, recognising that younger individuals generally have a lower risk of CVD. This could require you to adjust your hypothesis and possibly your RQ. Alternatively, you might think about including women in your study to broaden its scope.

Before beginning data collection, it is crucial to thoroughly consider your RQ and hypothesis. Research is iterative; you start with one or more RQs and a set of hypotheses, then design experiments to test these hypotheses. During this process, continually reassess whether your RQ and hypotheses are logical or if modifying them could make your project more compelling or impactful. Once you have the RQs and hypotheses with which you are satisfied, you can start designing your data collection strategy.

6.3 Survey Populations

Generally speaking, statistics is the science of drawing statements and conclusions about a population by using a sample of it (as we discussed in Section 5.1). Let us summarise some terminology here again so that you do not have to jump back and forth between chapters. The material in the following sections has been summarised and adapted from [1].

The term *population* normally refers to a set of objects (patients, molecules, galaxies, etc.) that are *infinite* (at least in theory) in nature and, due to this, cannot be known or described exactly. For example, the population of all results of tossing a coin is an infinite set of two possible results: head and tail. We do not have access to the infinite set, and thus we try, with statistics, to study and draw conclusions about it from a finite set of results. In reality there are no sets of objects that are infinite (for example all persons below 18 years of age that have lived so far on earth), but are large enough that it is impossible to know or describe. So, when defining populations, you can substitute the word *infinite* with the words *very large*.

We can now give an intuitive definition of survey sampling (or simply sampling) now.

Definition 6.3 (Survey Sampling) Survey sampling (or simply sampling) deals with the problem of selecting a finite set of elements from a potential infinite population.

The term survey refers to the act of collecting data.

Info Meaning of Survey

The word **survey**, according to the Britannica dictionary, refers to *an activity in which many people are asked a question or a series of questions to gather information about what most people do or think about something*. But the dictionary also gives the definition of *an act of studying something in order to make a judgement about it*. This second definition is much more apt and will serve the reader well. Practically, all data used in the scientific field are not coming from surveys but from measurements and experiments; thus, I much prefer the second definition.

Let us start with some formalism. The population, sometimes called the *universe* in sampling theory, is simply the set of N identifiers (we will assume that N is very large, as in reality there is no **infinite** collection of objects), such as, for example, patient names or the chemical composition of molecules. We can indicate the universe with $\mathbf{U}$:

$$\mathbf{U} = \{u_1, ..., u_N\} \tag{6.1}$$

Each identifier u_i represent a unique individual object in our population. Note that in this chapter, I use the terms population and universe with the same meaning. The population can be made up of anything, companies, people, land sections, chemical compounds, etc.

In our initial discussion, you may have got the impression that populations are given and simply exist. But even **defining** if an object is part of a population is not trivial. For example, suppose that you want to include in your population all smokers. How do you define if someone is a *smoker*? For example, a possible definition is [2] "*an adult who has smoked more than 100 cigarettes in his/her lifetime and currently smokes at least once a week*". Assessing whether a person is a smoker may not be as easy as it sounds. In sampling theory, you should always define what are called **eligibility criteria** that will define which object is part of the population and which not. This step is fundamental, but its relevance is particularly evident in medicine when selecting which patients should be in the population and which are not specific RQs in the medical context.

After designing your RQs and hypotheses, the next step is to define your population. In other words, you must design eligibility criteria that would define your **hypothetical** population. I have used the term hypothetical, meaning that you do not yet have data on all individual units in the population. Having the criteria allows you to decide whether, when presented with an object, this belongs to your population or not.

Info Research Project Design Process (Part I) (with Focus on Machine Learning and Data)

Let us start by giving the steps to setting up a machine learning project in a more concise form.

a) Write one (or multiple) **RQs**.
b) Design a set of **hypotheses** that will help answer your RQs in some way (recall that RQs are very generic in nature and allow for a wide array of possible answers).
c) Design the **population** (the data you will need to test your hypotheses) by defining eligibility criteria considering your RQs and hypotheses. Note that defining your population does not mean that you have (or will collect) data on all individual objects contained in the population. It means simply that you will be able to decide whether an object is contained or not in the population. Data collection will follow. After all, populations are typically very large, and it is impossible to get data for all objects contained in them.

Designing the experiments will follow sampling from your population (more on that later in the chapter). You are not done yet! After discussing sampling, we will discuss the next steps.

6.4 Survey Samples

A survey sample, denoted by **S** is simply a subset of the survey population (or universe) **U**

$$\mathbf{S} = \{u_{j_1}, ..., u_{j_m}\} \subseteq \mathbf{U} \tag{6.2}$$

where the $j_1, j_2, ...$ are integers. There are two main approaches to select a survey sample: *probability* and *non-probability* sampling.

6.4.1 Non-probability Sampling

Non-probability sampling simply means selecting elements from the survey population according to fixed rules and not by chance. Sometimes, a specific sampling strategy is chosen due to specific limitations (such as time or budget constraints). Here are some examples of non-probability sampling.

- **Restricted Sampling**: sampling is simply done keeping only parts of the population that are easily accessible. Maybe you are working in a hospital and thus your sample includes only patients from your hospital.
- **Judgement sampling**: sampling is obtained based on what the sampler believes to be *representative*. Maybe you are studying brain tumours and how they appear in MRI images. You may decide to study only specific types of tumour, as your experience has shown that, in general, they appear in MRI images similarly to most tumours.
- **Convenience sampling**: sampling is performed simply by keeping what is easily *reachable*. This type refers more to classical surveys, in which people had to reach people to ask questions.
- **Quota sampling**: the sample is gathered by several interviewers (for example, when talking about surveys), each tasked with collecting a specific quantity of units that possess particular types or characteristics. The selection of these units is entirely up to the discretion of the interviewers. If you are not dealing with interviews, you may have a certain number of people, each tasked with getting a certain number of objects you want to study (for example, you may have a certain number of chemists, each tasked with getting a certain number of chemicals for your study).

These methods are used when sampling by chance (see next section) is not feasible or simply too time-consuming or expensive. Statistical validity of the results with such samples relies **strongly** on assumptions. For example, in judgement sampling, the analysis relies on the assumption that the sample is representative, something not everyone may agree on. Typically, these methods are chosen because they are faster or less expensive. Non-probability sampling is a widely used method for gathering essential data on human populations, extensively employed by researchers in social and behavioural sciences, as well as those in medical and health fields. The volume of data required can be vast, encompassing tens or even hundreds of measurements per participant (blood samples, urine analysis, medical imaging, psychological assessments, etc.). For such studies, non-probability sampling might represent the only practical method.

The process works according to the following steps.

1. **Define the population**: determine who or what you want to study. Unlike probability sampling (more on that later), you do not need to have a complete list of the population (in other words, you do not need, for example, a list of names of all possible participants in your study). You just need criteria on how to define your population.
2. **Choose a non-probability sampling method**: select the most appropriate non-probability sampling technique based on your RQs and hypotheses and the nature of your population. Common methods we mentioned include convenience sampling, judgement sampling, etc.
3. **Determine sample size**: decide on the number of participants or objects you need. This decision may be influenced by factors such as the depth of analysis required, the time and the available resources (it may very well be a budget issue).

Non-probability sampling does not rely on statistical formulas for sample size determination and is typically determined by practical reasoning (like how much money you have, how much time, etc.).

4. **Recruit participants**: based on the chosen method, begin recruiting participants or getting your objects. For example, in convenience sampling, you collect data from individuals who are readily available. If you are studying chemical compounds, you would need to go and buy them for your study. You may decide to buy only compounds that are safe to use, or inexpensive to buy, for example.
5. **Collect data**: once your sample is selected, collect the data necessary for your study. This could involve surveys, interviews, observations, etc.

Remember, while non-probability sampling can be more practical or the only option available for certain studies, it may introduce bias and limit the generalisability of the results.

6.4.2 Probability Sampling

Probability sampling simply means that a sample is obtained by selecting elements of the population in a random fashion (more precisely according to some probability measure). Each element is given a probability of being selected (an easy approach is to give all elements of the population the same probability of being selected) to remove bias associated with subjective decisions (for example, if using judgement sampling). If you have a large population at your disposal, you can use this approach to randomly select a sample.

Info ★ Probability Sampling Design

Let us define

$$\Omega = \{\mathbf{S} | \mathbf{S} \subseteq \mathbf{U}\} \tag{6.3}$$

the set of all possible subsets of **U**. We can define $\mathcal{P}$ a probability measure over Ω such that

$$\mathcal{P} \geq 0 \ \forall \mathbf{S} \in \Omega \tag{6.4}$$

and

$$\sum_{\mathbf{S}:\mathbf{S} \in \Omega} \mathcal{P}(\mathbf{S}) = 1 \tag{6.5}$$

By using $\mathcal{P}$ you can then create your sample.
For example, your population may be all patients of a hospital in the year 2023. You may decide that $\mathcal{P}$ is equal to zero for all patients younger than 50 years of age when they entered the hospital and equal to $1/m$ for all other patients, where m is the number of patients older than 50 years of age who entered the hospital in 2023. By sampling using $\mathcal{P}$ your sample will contain only patients older than 50 years of age.

The process works according to the following steps.

1. **Define the population**: determine the entire group of individuals you want to study. This could be all students in a school, all employees in a company, etc. This comes from your RQ and hypotheses, as we discussed.
2. **Create a list**: compile a complete list of all members of the population. Each member is assigned a unique identifier, such as a number or string.
3. **Random selection**: use a random method, such as a random number generator, to select a specific number of individuals/objects from the list you created in the previous step. The number of individuals selected depends on the desired/possible sample size. Note that depending on your RQ or hypotheses, not all elements of the population must have the same probability of being chosen.
4. **Conduct the survey**: gather data from individuals or objects chosen by random selection. The sample size determined in the previous step is directly influenced by the data requirements. For example, if acquiring the information is costly, you might opt for a smaller sample size.

The random method you use in step 3 can be more complicated than using an equal probability for each individual/object in your population. That will depend on your RQ and the hypothesis you are studying.

6.5 Stratification and Clustering

A population **U** is stratified if it is divided into q non-overlapping groups (called **strata**):

$$\mathbf{U} = \mathbf{U}_1 \cup \mathbf{U}_n \cup \ldots \cup \mathbf{U}_q \tag{6.6}$$

For example, patients can be divided in subgroups each having a different disease, into different age groups, etc. A population is said to be clustered if it can be divided in subgroups (called **clusters**). The two definitions may seem exactly equivalent, but the difference lies in the way in which *clusters* and *strata* are used. When dealing with a stratified population, sampling involves selecting elements from **all** strata. When dealing with a clustered population, only a portion of the clusters will appear in the final sample. For example, if we stratify patients in different age groups, then when creating our sample we will select patients from **every** age group in the population. In contrast, if we cluster people geographically, our final sample may contain only a subset of regions that we have at our disposal.

In Chapter 7 you will learn that one of the easiest ways of checking how a model is generalising is to split your data into two parts (the so called hold-out approach). When doing this, if your population is stratified, be careful that both your *subsets* contain elements from all strata.

6.6 Random Sampling without Replacement

In this method, once an individual or object is selected from the population for inclusion in the sample you are creating, it cannot be chosen again. This approach ensures that each member of the population can be selected, but no individual or object can be included in the sample more than once. It is typically used when the goal is to avoid duplicating members in the sample, since your RQ or hypotheses require it. For example, if you are drawing cards from a deck, once a card is drawn, it is set aside and not put back into the deck for subsequent draws.

Warning Random Sampling without Replacement

Here is an overview of the advantages and disadvantages.

- **Advantages:** the method ensures each member of the population can be selected only once, preventing duplicates in the sample. This can lead to a more diverse and representative sample. Additionally, the maximum sample size is limited to the population size, making it easier to manage.
- **Disadvantages**: the method requires a comprehensive list of the population beforehand, which can be difficult or impractical to obtain for large or dynamic populations. In addition, managing and tracking selections from a very large population can be more complex and resource intensive.

6.7 Random Sampling with Replacement

In this method, after an individual or object is selected from the population and included in the sample, it is put back into the population, making it eligible for selection again in the next iteration. This method allows for the possibility of the same individual being put multiple times in the sample. It is particularly useful in simulations and bootstrap methods (see Section 6.8), where the objective is to create multiple independent samples from a single dataset to estimate the distribution of a parameter. An example of this would be to draw a card from a deck, noting its value, and then putting it back in the deck before drawing again. Note that with this approach, you can create a sample that is larger than the population from which you are sampling, since you can select an element multiple times. You should be very careful in doing that to avoid introducing bias in your sample that could skew your statistical results.

Warning Random Sampling with Replacement

Here is an overview of the advantages and disadvantages.

- **Advantages:** each selection is independent of the others, making the process simpler and often more suitable for theoretical or computational studies, like bootstrapping (more in Section 6.8). Furthermore, the method allows for a sample size that can exceed the population size, providing flexibility in experimental design and analysis. It can be particularly useful when the population size is small, as it allows a larger sample size without the constraint of exhausting the population.
- **Disadvantages**: when using this method there is the possibility of selecting the same individual or object multiple times, which can lead to duplicates in the sample, affecting diversity and potentially skewing results. Furthermore, the presence of duplicates might result in biased estimates of population parameters if not properly accounted for in the analysis. Finally, while it allows for greater sample sizes, it may result in a sample that is less representative of the population, especially if the population is large and diverse.

6.8 ■ Bootstrapping

Bootstrapping is a powerful sampling approach that does not rely on large sample sizes or strict assumptions about the population distribution. It involves repeatedly *resampling with replacement* from a population to create a large number of "bootstrap samples". These samples are then used to estimate the distribution of a statistic (mean, median, variance, etc.) by calculating the desired statistic on each of the bootstrap samples. This process allows for the estimation of the standard error, confidence intervals, and significance testing of the statistic, providing a way to understand the variability and reliability of the estimate from the original dataset.

Let us make an example to see how you could use bootstrapping to do a statistical study of a model accuracy (for a deeper understanding, you should read Chapter 7 on model validation). Suppose your dataset consists of 1,000 customer records. Each record includes features related to customer behaviour and a label indicating whether the customer churned (left the company or cancelled some kind of contract with the company). The first step would be to create bootstrap samples. You could, for example, create 1000 bootstrap samples from your original dataset. Each bootstrap sample is created by randomly selecting records with replacement, meaning that the same record can appear multiple times in a sample, and each sample is the same size as the original dataset. Then you should train a model on each of the bootstrap samples. We will assume that you have a validation dataset that was not used to generate bootstrap samples. You need to evaluate the performance of all 1000 models you trained on the 1000 bootstrap samples on the validation dataset. After performing the above steps for all bootstrap samples, you will have a distribution of accuracy scores for your model of which you can calculate the average (remember our discussions in Section 5.17 on how variance is influenced by sample size). Note that this method is a variation of what is called Monte-Carlo cross validation described in Chapter 7.

6.9 Random Stratified Sampling

If you have a stratified population, you should pay attention to having all strata in your sample. You can obtain this by using the following process, assuming that you have your population, but you have not yet stratified it. The following steps will help you obtain a stratified population and sample.

1. **Identify the stratifying variable**: choose the features you will use to divide the population into different strata and decide how to split your population in strata. This should be a characteristic that is believed to influence the outcome of the research, such as age, sex, income level, etc. In addition to deciding that, for example, age should be used for stratification, you also have to decide in which age groups you want to stratify your population. This is a two-step process: decide the features and the feature ranges that will define your strata.
2. **Divide the population into strata**: based on the stratification strategy from the previous point, divide the population into distinct strata. Each unit in the population should belong to one and **only** one stratum.
3. **Determine sample size**: decide on the total sample size for your study.
4. **Define the sample size for each strata**: determine how many individuals/objects to sample from each stratum. This can be done proportionally (*proportional allocation*) based on the size of the strata in the population or equally among the strata regardless of their size in the population or based on other considerations relevant to your RQs or hypotheses.
5. **Select samples from each stratum**: within each stratum, use random sampling to select individuals. This ensures that every member of the stratum has an equal chance of being included in the sample. You will select as many samples from each stratum as defined in the previous step.
6. **collect data**: proceed to collect data from selected individuals in all strata.

By following these steps, stratified random sampling allows you to obtain a sample that is more representative of the population (meaning statistically significant to the population), especially when there are significant differences between strata that could affect the study's outcomes. This method is often used when doing model validation, as in Chapter 7.

Consider an example to make this process and reasoning more concrete. Imagine a scenario in which you are trying to assess the impact of a new teaching method on student performance in mathematics across a region. This region includes a diverse array of schools, such as public and private, as well as urban and rural, each with varying levels of resources and socio-economic backgrounds among their students. The socio-economic status of students is a significant factor that can influence academic performance, with schools in rich areas typically having more resources and, consequently, potentially better student outcomes compared to schools in less wealthy areas.

To directly compare average scores across the entire region without accounting for these socio-economic differences could obscure the true effectiveness of the teaching method. This is because the method could perform differently in different environments: being more effective in some and less effective in others. Stratified sampling addresses this issue by ensuring that schools from each socio-economic category are represented in the sample. This allows for a more precise analysis of the teaching method's effectiveness across diverse socio-economic backgrounds.

In this context, you would need to first categorise the schools into different strata based on their socio-economic status (e.g., high, medium, low). Then you would need to decide on the sample size for each stratum to ensure proportional representation based on the number of schools or the student population within each socio-economic category. After this, schools and subsequently students within those schools are randomly selected from each stratum to participate in the study. The data on student performance in mathematics are then collected and analysed, with the analysis making comparisons both within and across the different socio-economic strata to assess the overall impact of the teaching method. Stratified sampling as described enables you to draw more accurate and generalisable conclusions regarding the effectiveness of the teaching method.

6.10 Sampling in Machine Learning

We have scratched the surface of sampling theory, which might lead you to believe that its application is limited only to statistical analysis. However, its significance extends to the realm of machine learning, particularly in the context of model validation. Note that model validation will be explored in detail in the following chapter (Chapter 7). If you have not had a chance to read it yet, please return to this section afterwards.

For model validation, it is necessary to divide your dataset, referred to as D, into two parts or sometimes several. For example, the hold-out method splits the dataset into two, whereas k-fold cross-validation divides it into k subgroups. This process of dividing your dataset is essentially a form of sampling from D. Let us examine how sampling theory is applied in machine learning and model validation through a systematic approach. Suppose that you have already defined a target population, established a sampling method, and collected the data, resulting in your data set D. It is important to apply sampling theory from the beginning to ensure that D is informative and capable of answering your research questions and testing your hypotheses. The application of sampling theory is crucial once more during the model validation stage. When you are splitting your dataset D in multiple parts, you always must ensure that each part is representative of the population. Here are some tips on how to do it.

1. If your population is stratified, use random stratified sampling as described in the previous section. The problem may be that often you do not know at the beginning how to define the strata. In this case define them as we have discussed in the previous section, by first identifying the stratifying variables and then divide the population into strata. Then you randomly select units from each strata with some strategy, such as, for example, with proportional allocation.
2. If your population is not stratified, simply use random sampling (with or without replacement) to decide which elements of the population are in which subgroups of the dataset D that you are creating.

To conclude, sampling theory is fundamental in creating a representative sample for machine learning experiments. The fowlloing info box outlines the steps necessary to design a research project (with a focus on machine learning).

Info Research Project Design Process (Part II) (with Focus on Machine Learning and Data)

Let us conclude the chapter by listing the steps to create a dataset efficiently.

a) Write one (or multiple) **RQs**.
b) Design a set of **hypotheses** that will help answer your RQs in some way (recall that RQs are very generic in nature and allow for a wide array of possible answers).
c) Design the **population** (the data you will need to test your hypotheses) by defining eligibility criteria considering your RQs and hypotheses. Note that defining your population does not mean that you have (or will collect) data on all individual objects contained in the population. It means simply that you will be able to decide whether an object is contained or not in the population. Data collection will follow. After all, populations are typically very large, and it is impossible to get data for all objects contained in them.
d) Create your **samples** using the sampling theory and methods that we have discussed in this chapter.
e) Get the data for the individual objects in your samples.
f) Design and carry out experiments to validate or disprove your hypotheses.

References

1. Changbao Wu and Mary E. Thompson. *Sampling Theory and Practice*. ICSA Book Series in Statistics. Springer International Publishing, Cham, 2020.
2. Changbao Wu, Mary E Thompson, Geoffrey T Fong, Qiang Li, Yuan Jiang, Yan Yang, and Guoze Feng. Methods of the international tobacco control (itc) china survey. *Tobacco Control*, 19(Suppl 2):i1–i5, 2010.

Chapter 7
Model Validation and Selection

> *In mathematics, the art of posing a question must be held of higher value than solving it.*
>
> Georg Cantor

This chapter addresses the fundamental aspects of model validation and selection in the field of machine learning. It begins by discussing the concept of model validation, emphasising its critical role in assessing a model's ability to generalise to new, unseen data. The phenomena of overfitting and underfitting are explored, along with an in-depth discussion on the bias-variance trade-off. The chapter then discuss various cross-validation techniques, including the hold-out approach, k-fold cross-validation, the leave-one-out method, and Monte Carlo cross-validation, each catering to specific scenarios and data characteristics. The selection of appropriate validation methods is discussed in the context of supervised and unsupervised learning models. Furthermore, the chapter delves into the complexities of model selection, highlighting the importance of balancing quantitative measures with qualitative criteria such as interpretability, domain relevance, and ethical considerations. This comprehensive exploration aims to equip practitioners with the knowledge and tools necessary for effective model evaluation and selection, ensuring the development of robust, reliable, and ethically sound machine learning models.

7.1 Introduction

Suppose that you trained a machine learning model to predict something (for a classification or regression problem). At a certain point, you will want to use the model on new data that you did not have during the training phase. Maybe you developed a model to predict some medical condition that hospital doctors need to use on new patients. Perhaps you have developed algorithms for a self-driving drone that needs to navigate new environments that you did not have in your training data. In any case, it is imperative to assess how the model will perform in those situations. In other words, you will need new techniques to assess what is called model *generalisation*. This simply means answering the question: how will the model perform in new situations? Will it continue to perform as it did on the training data or will its performance decline?

U. Michelucci, *Fundamental Mathematical Concepts for Machine Learning in Science*,
https://doi.org/10.1007/978-3-031-56431-4_7

This assessment is known as *Model Validation* (MV). MV encompasses all methodologies to verify that the model performs as intended, particularly in terms of its effectiveness, on previously unseen data. MV can be formulated in a slightly more general form as follows.

Definition 7.1 (Model Validation) Model validation refers to the process of confirming that a machine learning model actually achieves its intended purpose.

What exactly constitutes the "intended purpose" of a model? Typically, a model's main objective is to detect patterns within the dataset and ensure stable performance on new, yet statistically comparable, data to what was used during training. Put more straightforwardly, the effectiveness of the model should not significantly fluctuate as it deals with unfamiliar (not seen before) data. But in general, why should this be relevant? There are a few reasons for this.

- **Assessing Generalisation Properties**: validation helps in evaluating how well a model performs on unseen data. This is essential for understanding the model's ability to generalise beyond the training dataset (as we mentioned).
- **Detecting Overfitting**: overfitting occurs when a model learns noise and fluctuations in the training data to such an extent that it negatively impacts the performance on new data. Validation helps in detecting overfitting by providing a check on the model's performance on a separate dataset that was not used during training (this will be further explored later in the chapter).
- **Model Selection**: through validation, different models and their parameters can be compared and fine-tuned. This process allows for the selection of the best model and parameter set that offers the optimal balance between bias and variance (see the next sections for a discussion).
- **Trust**: in practical applications, the validation process is crucial for stakeholders to trust the model. Demonstrating robust performance ensures that the model is reliable and effective in real-world scenarios.
- **Regulatory Requirements**: in certain industries, such as finance and healthcare, model validation is not just good practice, but also a regulatory requirement to ensure that models are fair, transparent and reliable.

The focus of this chapter is to explain and discuss several techniques to achieve the above-mentioned goals. This is probably one of the most important parts of any machine learning pipeline in any project.

7.2 Bias-Variance Tradeoff

Before starting we need to clarify how results from a machine learning model depends from its (*not well defined*) complexity. Imagine the following scenario. You have an infinite (or at least very large) dataset of observations at your disposal. You have trained a model $\mathcal{M}$ with a subset D_T of the data and are testing your model on a different subset of the dataset D_V. Imagine that you are tracking the error of your model (it could be $1 - a$, with a accuracy, or the mean squared error MSE) on the two different datasets D_T and D_V. If the model is very simple (imagine a linear regression), both the errors on D_T and D_V will be high. It is easy to understand why. The simplicity of the algorithm will make it impossible for it to learn complex patterns that may be present in the data. By increasing the model's complexity, it is to be expected that both errors will go down, since the model increased flexibility will allow it to learn more finer structures in the data. But after further increasing the complexity of the model (imagine a very large neural network), what can (and will) happen is that while the error on D_T will go down (the model will learn to model all small details of the data it has been trained on), its error on new data (on D_V) will go up, since the finer details in D_T (that the model learnt) will be different from the finer details in D_V. This type of behaviour is common to all kinds of data and situations, so much so that those two different scenarios described are given specific names.

- **High-Bias Low-Variance**: when the model is very simple (imagine linear regression), it is said that we are in the High-Bias Low-Variance region. High-Bias since the model is performing very poorly, and Low-Variance since the results do not (more on this later, for now an intuitive understanding will suffice) vary too much (the results of the model trained on different datasets will not vary much).
- **Low-Bias High-Variance**: when the model is highly complex and flexible (imagine a large neural network), it is said that we are in the Low-Bias High-Variance region. Low-Bias since the model gets very low error values, but High-Variance since typically models are quite unstable, as they are learning finer and finer details in the data that vary substantially from the one in unseen data used for testing.

Typically, achieving a state of low bias and low variance simultaneously, which is considered the ultimate goal in machine learning, is not feasible. The most effective strategy involves striking a balance between these two aspects, a concept known as the **Bias-Variance Tradeoff**. This principle will be discussed in depth in subsequent sections. Specifically, Section 7.3 will provide a mathematical perspective to complement the intuitive overview provided so far. Following that, detailed discussions and examples of the two regimes will be presented to enhance the reader's comprehension.

7.3 ★ Bias-Variance Tradeoff - a Mathematical Discussion

Let us consider a dataset $D = \{(x_1, y_1), ..., (x_n, y_n)\}$ drawn i.i.d.[1] from some distribution $P(X, Y)$ where with the big letter X and Y we have indicated the random variables of which x_i and y_i with $i = 1, ..., n$ are the observed values in the dataset respectively. Imagine that we could have an infinite number of datasets.

[1] Remember that i.i.d. means **i**ndependent and **i**dentically **d**istributed.

Note that for a given x there may be multiple possible values of y. If, for example, x indicates the characteristics of an apartment (like the number of rooms, the number of square metres, etc.) and y its price, two apartments with the same characteristics can be sold at different prices. To continue the discussion, we will need some formalism. Let us start with the following. We will indicate by $\overline{y}(x)$ the expected label.

$$\overline{y}(x) \equiv \mathbb{E}_{y|x}[Y] \tag{7.1}$$

where the expected value is over all possible values of y given a certain x as input. We can also define an *expected model*

$$\overline{h}(x) = \mathbb{E}_D[h(x)] \tag{7.2}$$

that is nothing else as the model (the mathematical function $h(x)$) averaged over all possible datasets (the subscript D of the expected value operator $\mathbb{E}$). In other words, $\overline{h}(x)$ is the model obtained by averaging all the models trained on all possible datasets.

We are interested in studying the expected generalisation error ξ over all possible inputs x, output y and datasets D[2]. We will define the expected generalisation error as

$$\xi = \mathbb{E}_{x,y,D}[(h(x) - y(x))^2] \tag{7.3}$$

where $y(x)$ is the correct label for a given input x (imagine for the sake of this discussion that we deal with a regression problem). Basically, ξ is nothing more than the MSE evaluated on all possible inputs x, output y and datasets. We can now work on the expression to get more insights.

$$\begin{aligned}
\xi &= \mathbb{E}_{x,y,D}[(h(x) - y(x))^2] = \mathbb{E}_{x,y,D}[[(h(x) - \overline{h}(x)) + (\overline{h}(x) - y(x))]^2] = \\
&= \underbrace{\mathbb{E}_{x,y,D}[(h(x) - \overline{h}(x))^2]}_{A} + \underbrace{2\mathbb{E}_{x,y,D}[(h(x) - \overline{h}(x))^2(\overline{h}(x) - y(x))^2]}_{B} + \\
&+ \underbrace{\mathbb{E}_{x,y,D}[(\overline{h}(x) - y(x))^2]}_{C}
\end{aligned} \tag{7.4}$$

First of all note that C does not depend on which dataset the model has been trained on, thus

$$C = \mathbb{E}_{x,y,D}[(\overline{h}(x) - y(x))^2] = \mathbb{E}_{x,y}[(\overline{h}(x) - y(x))^2]. \tag{7.5}$$

Also, A does not depend on y thus

$$A = \mathbb{E}_{x,y,D}[(h(x) - \overline{h}(x))^2] = \mathbb{E}_{x,D}[(h(x) - \overline{h}(x))^2]. \tag{7.6}$$

Now let us prove that $B = 0$.

[2] This is where the topics described in the chapter on statistics comes in handy.

$$
\begin{aligned}
B &= \mathbb{E}_{x,y}[\mathbb{E}_D[h(x) - \overline{h}(x)](\overline{h}(x) - y(x))] = \\
&= \mathbb{E}_{x,y}[(\mathbb{E}_D[h(x)] - \overline{h}(x))(\overline{h}(x) - y(x))]] = \\
&= \mathbb{E}_{x,y}[(\overline{h}(x) - \overline{h}(x))(\overline{h}(x) - y(x))] = 0.
\end{aligned}
\tag{7.7}
$$

So we can write

$$
\xi = A + C \tag{7.8}
$$

Now let's work on C.

$$
\begin{aligned}
C &= \mathbb{E}_{x,y}[[(\overline{h}(x) - \overline{y}(x)) + (\overline{y}(x) - y(x))]^2] \\
&= \underbrace{\mathbb{E}_{x,y}[(\overline{y}(x) - y(x))^2]}_{\text{Noise}} + \mathbb{E}_x[\underbrace{(\overline{h}(x) - \overline{y}(x))^2}_{\text{Does not depend on } y}] + \\
&= \underbrace{\mathbb{E}_{x,y}[(\overline{h}(x) - \overline{y}(x))(\overline{y}(x) - y(x))]}_{D}
\end{aligned}
\tag{7.9}
$$

one can easily, as we have done above, prove that $D = 0$. Thus we can finally write

$$
\begin{aligned}
\xi = &\underbrace{\mathbb{E}_{x,D}[(h(x) - y(x))^2]}_{\substack{\textbf{Variance}\text{: capture how much your} \\ \text{model changes if you train it} \\ \text{on a different dataset}}} + \\
+ &\underbrace{\mathbb{E}_{x,y}[(\overline{y}(x) - y(x))^2]}_{\substack{\textbf{Noise}\text{: intrinsic noise due to} \\ \text{measurement errors, etc.} \\ \text{It does not depend on the model}}} + \\
+ &\underbrace{\mathbb{E}_x[(\overline{h}(x) - \overline{y}(x))^2]}_{\substack{\textbf{Bias}\text{:} \\ \text{the error of your} \\ \text{model even on } \underline{\text{an}} \text{ infinite dataset.} \\ \text{Remember that } \overline{h}(x) \text{ is the average} \\ \text{model over all possible datasets.}}}
\end{aligned}
\tag{7.10}
$$

This means that ξ has three contributions. Two of which do not depend on the dataset used to train the model.

Figure 7.1 shows the behaviour of variance, noise, and bias (the three components of the generalisation error) with respect to the complexity of the model in a typical case. As the complexity of the model increases, the bias decreases, as the model is able to capture finer details of the data. At the same time, the variance grows since the model will capture features of the data that are only dependent on the data used for training (on the model internal characteristics). When the model complexity is low, the bias is high because the model is not able to capture many details of the data, while the variance is low since the model is too simple to capture fine details.

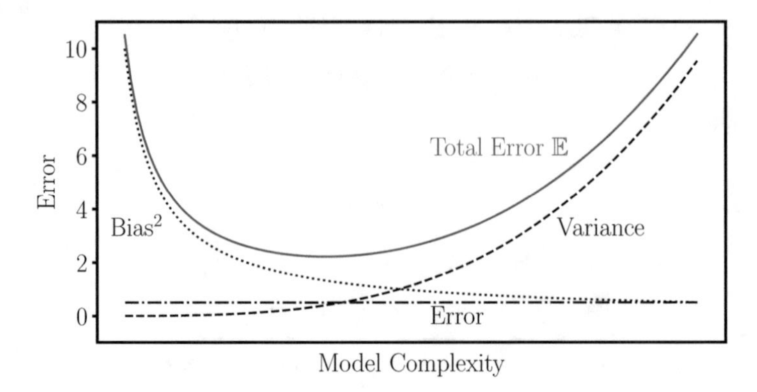

Fig. 7.1 The behaviour of the variance, Noise and Bias with respect to the model complexity. As you increase model complexity, the bias decreases, as the model is able to capture finer details of the data. At the same time, the variance grows since the model will capture features of the data that are dependent on the data used for training. When the model complexity is low, the bias is high, as the model is not able to capture much detail of the data, while the variance is low since the model is too simple to capture fine details.

Info ★Bias

In Equation (7.10) the last term

$$\mathbb{E}_x[(\overline{h}(x) - \overline{y}(x))^2] \tag{7.11}$$

is called Bias. Let us discuss why. Consider n random variables X_i for $i = 1, ..., n$, where X_i are i.i.d. all with expected value μ and variance σ^2. Let us define the quantities

$$\overline{X} = \frac{1}{n}\sum_{i=1}^{n} X_i \tag{7.12}$$

and

$$S^2 = \frac{1}{n}\sum_{i=1}^{n}(X_i - \overline{X})^2 \tag{7.13}$$

Let us do some calculations.

$$
\begin{aligned}
\mathbb{E}[S^2] &= \mathbb{E}\left[\frac{1}{n}\sum_{i=1}^{n}(X_i - \overline{X})^2\right] = \mathbb{E}\left[\frac{1}{n}\sum_{i=1}^{n}\left((X_i - \mu) - (\overline{X} - \mu)\right)^2\right] \\
&= \mathbb{E}\left[\frac{1}{n}\sum_{i=1}^{n}\left((X_i - \mu)^2 - 2(X_i - \mu)(\overline{X} - \mu) + (\overline{X} - \mu)^2\right)\right] \\
&= \mathbb{E}\left[\frac{1}{n}\sum_{i=1}^{n}(X_i - \mu)^2 - \frac{2}{n}(\overline{X} - \mu)\sum_{i=1}^{n}(X_i - \mu) + \frac{1}{n}(\overline{X} - \mu)^2\sum_{i=1}^{n}1\right] \\
&= \mathbb{E}\left[\frac{1}{n}\sum_{i=1}^{n}(X_i - \mu)^2 - \frac{2}{n}(\overline{X} - \mu)\sum_{i=1}^{n}(X_i - \mu) + \frac{1}{n}(\overline{X} - \mu)^2 n\right]
\end{aligned}
\tag{7.14}
$$

This can be further simplified with relatively easy steps to obtain

$$
\mathbb{E}[S^2] = \frac{1}{n}\sum_{i=1}^{n}(X_i - \mu)^2 - \mathbb{E}\left[(\overline{X} - \mu)^2\right] \tag{7.15}
$$

That is nothing else as

$$
\mathbb{E}[S^2] = \sigma^2 - \mathbb{E}\left[(\overline{X} - \mu)^2\right] \tag{7.16}
$$

The second term in Equation (7.16) is what is called *bias*, and is the equivalent of the last term in Equation (7.10), where the symbols correspondence is given by

$$
\overline{X} \leftrightarrow \overline{h}(x) \quad , \quad \mu \leftrightarrow \overline{y}(x) \tag{7.17}
$$

Since in Equation (7.16) the term

$$
\mathbb{E}\left[(\overline{X} - \mu)^2\right] \tag{7.18}
$$

is subtracted from the variance, that is the standard deviation squared, sometimes it is referred to as the Bias^2 (bias squared).

In general, you cannot have low bias and low variance at the same time. You always have to find a compromise between bias and variance (you can do this by choosing the right model, and thus its capacity of learning finer or rougher features of the data). This is what is called the **bias-variance trade-off** (as mentioned in the introduction of this chapter).

One of the key questions to answer while training a model is which regime you are in: low bias-high variance, high bias-low variance, or exactly at the sweet spot (the point in Figure 7.1 with the lowest total error)? In general, it is not possible to exactly calculate the different components of the total error, but there are methods that will help you to understand in which regime you are without the need to exactly evaluate the terms in Equation (7.10).

7.4 High-Variance Low-Bias regime

When you are in the *High-Variance Low-Bias* regime, training a model with different data will give very different and inconsistent models. To better understand this, let us study an example generating some data according to the following steps.

1. We generate an array of points $p_i = (x_i, y_i)$ for $i = 1, ..., 20$ with $x_i = 1/2(i-1)$ and $y_i = \cos(x_i) + 0.3x_i$.
2. We generate a second array of points p_i^n by adding random noise between -0.5 and 0.5 to y_i.

Now let's fit the data with a 20th degree polynomial, the result of which will be indicated with $p_{21}(x)$[3]. In Figure 7.2 you can see a plot of p_i^n, of the function $\cos(x) + 0.3x$ and of the polynomial fitted. It is very clear from Figure 7.2 that the

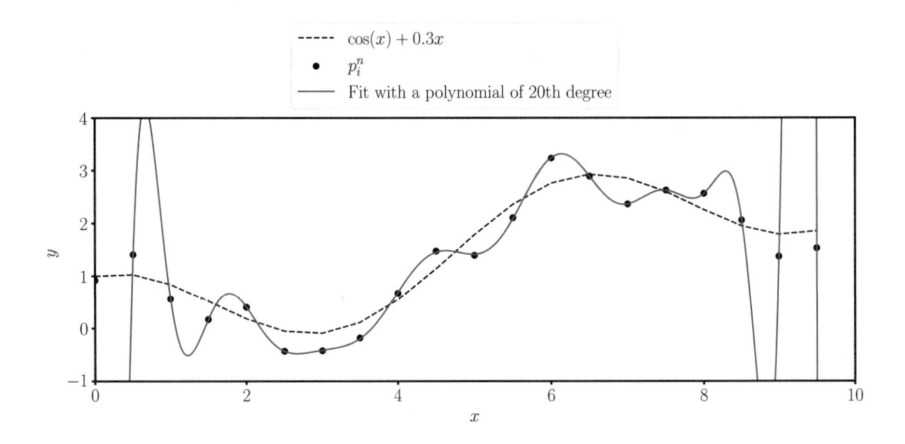

Fig. 7.2 A plot of the array of points p_i^n, of the function $\cos(x) + 0.3x$ used to generate the data without noise, and the result (in red) of fitting a 20th polynomial to the data with noise.

red curve (the fitted polynomial) is very different from the dashed line (the function used to generate the data). The polynomial has learnt to model the noise. Let us now do a small experiment. Let us generate multiple sets of points $p_i^{n,j}$ with $j = 1, ..., 10$ where each time (each j) we add new random noise to our function. Then let us again fit a 20th degree polynomial to the multiple set of points. In Figure 7.3 you can see the result. You can see what substantial difference the results are in Figure 7.3. The models are very different. This is not what we want, as this tells us that simply using different data for training will lead to substantially different results, or in other words to bad generalisation properties.

[3] The reason of the subscript 21 is that a 20th degree polynomial has also a constant term, so effectively 21 terms.

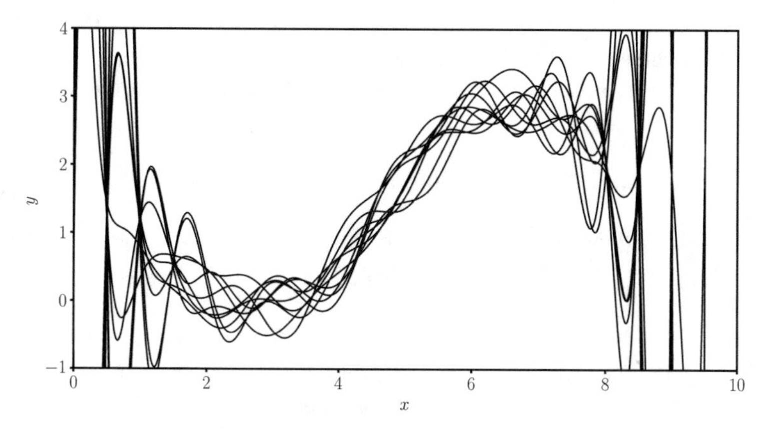

Fig. 7.3 The result of fitting a 20th polynomial to multiple dataset of points obtained by generating each time different random noise.

Consider a model (trained polynomial) trained on a given dataset D_T. We can now generate 10000 additional datasets simply by generating each time new random noise and evaluate the MSE between the model output and the 10.000 datasets. In Figure 7.4 you can see the distribution of the MSE. The dashed line is the MSE value obtained with the training dataset D_T. How should we interpret Figure 7.4?

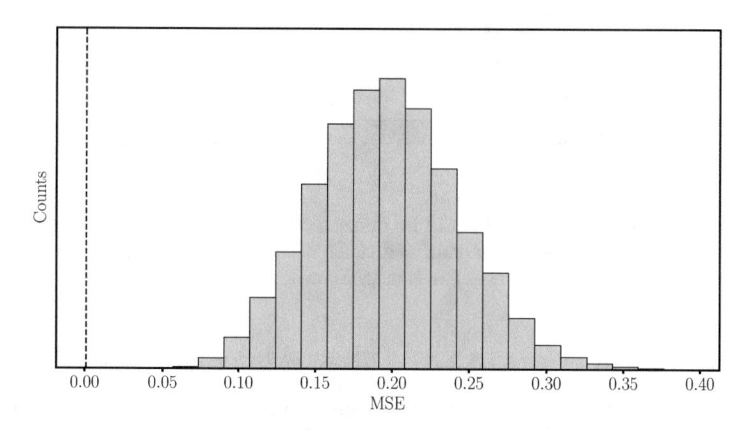

Fig. 7.4 Distribution of the MSE obtained by fitting a 20th polynomial to one dataset, and then evaluating the MSE between the polynomial and 10000 datasets obtained by generating each time new random noise. The dashed vertical line is the MSE evaluated on the training dataset.

It is clear that the model works very well on the training data D_T (the dashed line indicates a value of the MSE very close to zero), but has higher and wildly varying results when applied to new data (the histogram indicates the distribution of such results). The MSE ranges from about 0.05 to 0.35. This regime (High Variance and Low Bias) is characterised typically by performances of models that are extremely good on the training set but relatively bad on new data, unseen during the training phase.

Info High-Variance Low-Bias Regime

This regime is typically characterised by models that are extremely good on the training set, but significantly worse when applied to new data that were not seen during the training phase.

What happens when we move toward regimes of lower variance and higher bias? To study that in our example, it is enough to reduce the complexity of our model. We can do that by simply using a lower-order polynomial. In Figure 7.5 you can see the results for a 6th polynomial obtained the same way as in Figure 7.4. Now it is

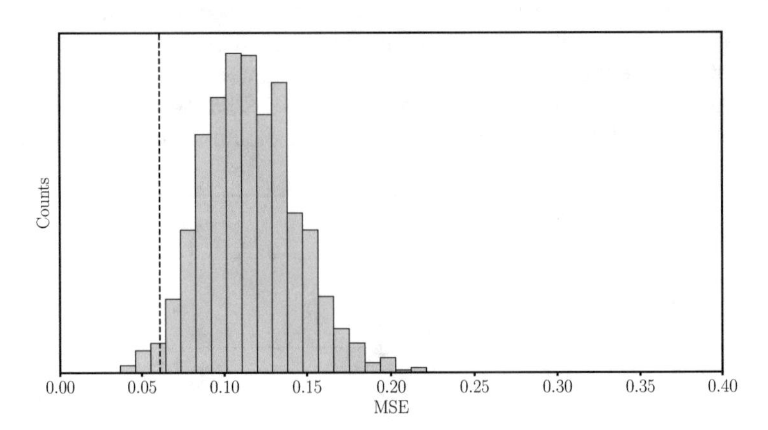

Fig. 7.5 Distribution of the MSE obtained by fitting a 6th polynomial to one dataset, and then evaluating the MSE between the polynomial and 10000 datasets obtained by generating each time new random noise. The dashed vertical line is the MSE evaluated on the training dataset.

important to compare Figures 7.4 and 7.5. There are two very important differences:

- The first is the fact that the MSE evaluated on the training dataset is not zero anymore but is moving toward the average of the MSE obtained on the new data (the histogram).
- The second important difference is that the spread of the distribution (its variance) is much smaller, indicating that the model, when applied to unseen data, is much more stable than the one shown in Figure 7.4. Note that this is not always the case and, very often when going to high-bias regimes, the average MSE may (and

will) grow again since the models will be less and less efficient in learning the data characteristics.

The difference between the MSE evaluated on the training data and the one on the new data is the metric that we can use to understand in which regime we are in. To convince you further, let us quickly check what happens when we move to low-bias regimes and then summarise everything.

7.5 Low-Variance High-Bias regime

To simulate a low-variance high-bias regime, let us consider a simple linear model. It is clearly too simple to capture the characteristics of the data. You can see in Figure 7.6 the results obtained analogously to Figure 7.3 with a linear model. The

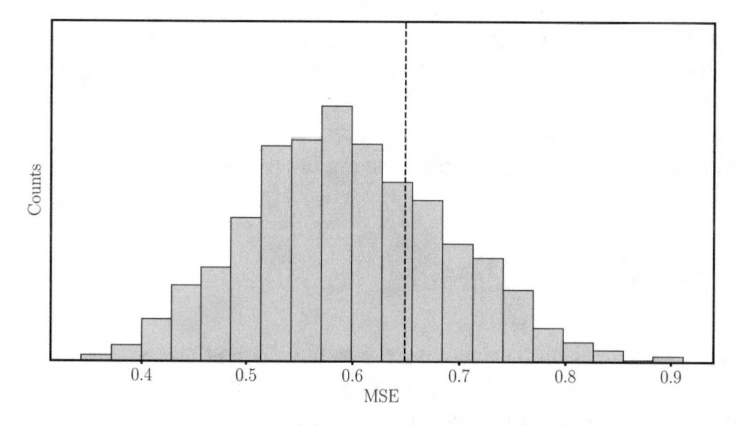

Fig. 7.6 Distribution of the MSE obtained by fitting a linear model to one dataset, and then evaluating the MSE between the polynomial and 10000 datasets obtained by generating each time new random noise. The dashed vertical line is the MSE evaluated on the training dataset.

differences with the previous examples are striking. The main one is the fact that the MSE evaluated on the training dataset is very close to the average MSE obtained with unseen data. Also note that the absolute values of the MSE are much higher than those obtained with the higher order polynomial (as we discussed in Figure 7.1). In this regime, the performance evaluated on the training data is comparable to that obtained on unseen data. In general, the models in this regime do not work well and are often not usable.

Info How to Determine the Regime in which a Model operates

To determine in which regime a trained model is, you can use the following criteria.

1. The model is in the **High-Variance Low-Bias** regime when the performance obtained with the training data is much better than the one obtained with unseen data (right side in Figure 7.1).
2. The model is in the **Low-Variance High-Bias** when the performance obtained with the training data is of the same order as the one obtained with unseen data and its value typically indicates a (very) poor performance (left side in Figure 7.1).
3. The model is in the ideal regime when the performance obtained with the training data is of the same order as that obtained with unseen data and its value indicates a good performance (at the point where the total error has a minimum in Figure 7.1).

The reader may be confused at this point and ask how we can estimate the performance of a model on "unseen" data. What is meant exactly? In our example, we generated thousands of different datasets to perform our tests, but in reality, data cannot be (at least not typically) generated in an arbitrary large quantity. How can we use the method discussed above in real situations? It is easier than you might think, and the most used approaches are discussed in the following sections.

7.6 Overfitting and Underfitting

There is an additional discussion we need to have before proceeding to MV techniques. It is necessary to define some terminology that is typically used in the machine learning community: **overfitting** and **underfitting**.

Definition 7.2 (Overfitting) Overfitting in machine learning is a phenomenon in which a model learns the training data too well, including its noise and random fluctuations, rather than just learning the underlying patterns. As a result, while the model may perform exceptionally well on the training data, its performance deteriorates significantly on new, unseen data. This happens because the model becomes too complex, capturing peculiarities of the training data that do not generalise to other data sets. In essence, overfitting involves the model memorising rather than learning to generalise from the training data. One says that the model is *overfitting* when in the high-variance low-bias regime.

Definition 7.3 (Underfitting) Underfitting in machine learning occurs when a model is too simple to capture the underlying patterns and complexities in the data. This typically happens when the model does not have enough capacity (in terms of parameters or structure) or when it has not been trained sufficiently long. As a result, the model performs poorly even on the training data, failing to establish the fundamental relationships within it. Underfitting indicates that the model has not learnt enough from the training data and, as such, will also perform inadequately on new, unseen data. It is essentially the opposite of overfitting, where a model is excessively complex; underfitting is due to an overly simplistic model. One says that the model is *underfitting* when in the high-bias low-variance regime.

Overfitting occurs in the right part of Figure 7.1 (high-variance low-bias), while underfitting in the left part of Figure 7.1 (low-variance high-bias). For our discussion we typically use the terms **overfitting** and **underfitting** to indicate the two regimes.

7.7 The Simple Split Approach (a.k.a. Hold-out Approach)

As discussed multiple times now, it should be clear that we need *unseen* (not seen during the training process) data to be able to validate our model. In the above examples, we simply generated new data, but in any real-life situation, this is not possible. The easiest way to find *unseen* data (that were not used for model training) is to take a portion of the available data and put it aside before training a model. Granted, we will have less data for training, but we will at least be able to validate our model. This approach is called *hold-out approach* since we keep a portion of the dataset (the hold-out set) to validate our model. This approach works according to the following steps.

- We split by using random sampling the datasets D in two portions: training (D_T) and validation (D_V)[4].
- We train the model on D_T and evaluate its performance on D_V.
- We compare the metrics chosen evaluated on D_T and D_V. If the two results are comparable, then we can deduce that the model seems to generalise well (recall that the model has not seen the D_V dataset). If the model performs much better on D_T than on D_V, then we are clearly in the overfitting regime. In this case, we need to choose a less complex model, or reduce its complexity if possible (especially for neural networks, regularisation try to achieve exactly this, that is, reduce the network complexity).
- If the metrics evaluated on D_T and D_V are similar, try to increase the complexity of the model to ensure that you are not in the low-variance high-bias regime. As soon as the difference between the metrics starts to rise, you are now entering the overfitting regime.

Warning What to consider in the Hold-out approach

The Hold-out approach seems simple enough, but it hides some challenges that you should be aware of.

[4] Note that sometime this portion of data is also called *test* dataset.

- Let us suppose you are trying to solve a classification problem. You should always check, after splitting the dataset, that all the classes in D_T and D_V appear in the same proportions. What happens if the dataset is sorted? Consider the MNIST [a]dataset and imagine that the first 600 images are all of digit 0, the next 600 of digit 1 and so on. If you split the data simply by taking the first 80% and the last 20%, you will get D_T and D_V with different classes in them, and thus you will have a biased training (your model will only see some digits) and a useless validation. Always check that the two portions are similar in the statistical sense.
- How big should D_T and D_V be? A typical choice (for reasons unknown to the author and perhaps based loosely on the Pareto principle[b]) is 80% for D_T and 20% for D_V. But this is not a strict rule. You will always have to find a good balance between having enough data for training (a sufficiently large D_T) and a D_V that is representative of your problem. In D_V you want to have as many representative examples as possible to be able to judge the model's generalisation properties realistically. When dealing with very large datasets, typical proportions between D_T and D_V are 90%/10% or even 99%/1%.

[a] As a reminder, MNIST, an acronym for the Modified National Institute of Standards and Technology database, is a large dataset commonly used in the field of machine learning. It comprises 70,000 handwritten digits (0 through 9), divided into a training set of 60,000 images and a test set of 10,000 images. Each digit is contained within a 28x28 pixel grayscale image.

[b] Wikipedia has more information on the principle at `https://en.wikipedia.org/wiki/Pareto_principle`.

7.8 Data Leakage

The hold-out approach seems simple enough, but there are many common mistakes that can be made. The most important will result in what is called *data leakage*. Data leakage in machine learning and data science refers to a situation in which information from outside the training dataset (for example, from the test dataset) is inadvertently used to create or train the model. This leakage of information typically results in a model that performs unrealistically well. Data leakage can occur in various forms. A typical example worth mentioning is that if data preprocessing (such as normalisation, feature selection, or filling missing values) is done using the entire dataset rather than separately on the training and test sets, information from the test set can inadvertently influence model training. Suppose that you want to normalise some feature of your data to have an average of zero. You can simply subtract the average from the feature value, but how will you calculate the average? Only from the training data or from the entire dataset before splitting it (suppose that you are using the hold-out approach)? If you are calculating the average from the entire dataset, some information from the test dataset will influence the training dataset, and thus the model training. Common examples of leakage are summarised below.

1. **Preprocessing Leakage**: If data preprocessing (such as normalisation and feature selection) uses the entire dataset rather than just the training set, information from the test set can leak into the model.
2. **Model Selection Leakage**: When model selection or hyper-parameter tuning is done using the test set, it may lead to an overfitting on the test data, providing a misleadingly high performance estimate (more on that later).
3. **Temporal Data Leakage**: In time-series data, using future data in the training process leads to leakage, as the model gets access to information it would not have in a real-world scenario.

To prevent data leakage in hold-out cross-validation, the following steps can be taken:

- **Preprocessing within Splits**: ensure that all preprocessing steps are conducted within each dataset portion (e.g., in the hold-out approach in the training and test portions). This means applying transformations after the split and separately to each portion.
- **Do not use any test data**: do not use any information from the test dataset (as averages, standard deviations, missing data or imputation approaches) on the training dataset.

Data leakage in hold-out cross-validation can lead to overly optimistic and misleading results.

Info Normalisation without Data Leakage

There are two typical approaches to normalise continuous features in a dataset.

- **Max-Min Scaler**: in other words modifying a feature for it to go from zero to one. To achieve this, a generic feature X is analysed and transformed into a new Y with

$$Y = \frac{X - \min(X)}{\max(X) - \min(X)} \tag{7.19}$$

If you decide to use this approach, the typical way of using it is to apply it separately on the training and test dataset (assuming that you are using the hold-out approach). In some instances, you can get the maximum and minimum from the training dataset and also apply the above formula to the test dataset. The only danger in doing this is if the data in the two datasets are significantly different. For example, you may have larger values in the test dataset for your feature that would result in a normalised feature value larger than one. This may pose problems depending on the model you are using, but in most cases this factor can be neglected.

But you could also use a slightly modified and more informed approach. Suppose that you know from your problem settings that your feature will have some maximum and minimum (for example, you may decide that the maximum age for patients will be 100 years, regardless of the data, or maybe you are studying children and thus decide that the maximum weight will be 50 kg). In this case, you may decide the values of the maximum and minimum independently of the data to the known limits. Then, you could apply the same transformation to both the training and test dataset without fear of data leakage.
One notable problem with this approach is that it does not remove the effect of outliers, it simply rescales them in a fixed range. So if your feature has large outliers, rescaling it will not solve the problem, it will simply translate it in different numerical range.

- **Zero Average-Standard Deviation of one**: Zero Average-Standard Deviation of One Normalisation, commonly known as standardisation (this is the typical method people use, so if you read just *standardisation* the change are very high that they are referring to this one), or Z-score normalisation, is a widely used data preprocessing technique in machine learning. The method involves adjusting the scale of each feature in the dataset so that they have a mean of zero and a standard deviation of one. The formula for this transformation is given by

$$Y = \frac{(X - \mu)}{\sigma} \tag{7.20}$$

where X is the original feature value, μ is the mean of the feature, and σ is the standard deviation of the feature. The same considerations that we discussed for the Min-Max scaler apply to this case.
In general, standardisation is beneficial for models sensitive to the scale of input features, like SVMs, Neural Networks, and models employing gradient descent. This method ensures that all features are on the same scale, aiding in model consistency, especially when features have different units or ranges.

The main message you should take out of this section is that splitting the data set brings with it a number of challenges and possible pitfalls. You should be aware of them and take care in doing things properly.

7.8.1 Data Leakage with Correlated Observations

Data leakage can manifest itself in subtle ways that are often difficult to detect. Consider a scenario in which you are working with a dataset comprising n patients, each represented by m MRI images. This results in a total of nm observations. It could be tempting to simply shuffle this dataset and then divide it, especially if a hold-out strategy is used for validation. However, this approach can introduce a significant source of leakage. Since MRI scans of the same patient are inherently correlated, reflecting the same medical conditions, having images of an individual patient distributed across both training and validation sets can lead to overly optimistic estimates of your model performance. This is because the presence of highly correlated data in both sets does not accurately reflect the model ability to generalise to unseen data. If you are in such a situation, you should split the dataset on a patient level, not on individual observations. This kind of scenario happens very often, and too often the dataset split is done incorrectly. Your situation may be different, you may have instead of patients, mechanical pieces and instead of MRIs multiple measurements for each machine, but the idea is the same. Split your dataset based on the machines, not on individual measurements.

Let us try to formalise this a bit more. If, in your dataset, you have subgroups of observations that are highly correlated for some reason, you should not split those subgroups and keep them in only one dataset: training or validation.

7.8.2 Stratified Sampling

As mentioned multiple times, you should pay attention to the splitting of the dataset. Let us give an example. Suppose that we have two classes, 0 and 1 and that we have 1000 inputs. Imagine that the data are not shuffled and that the first 500 labels are 0 and the second 500 are 1. If we split the data 80% for D_T and 20% for D_V simply by taking the first 800 and then the remaining 200 of the dataset, D_V will contain only the label 1. In this case, we will never be able to properly assess the performance of a model, since we will have no information on how the model performs on class 0 on unseen data (the D_V). This would not happen if the dataset was shuffled to start with, but this is not guaranteed in real life cases. The situation is even more complicated if we have multiple classes. A general rule of thumb is to always check how classes are distributed in the training and validation datasets. To split the data considering these aspects, we need to use a more sophisticated approach, called *stratified sampling*.

Warning **What Could go Wrong?**

When using the hold-out approach, you should be careful of situations where in splitting the dataset you end up with a D_T that is not representative of the original dataset (or the original research question). The problem being that you will train your model on D_T and thus on a different statistical distribution of the inputs. To better understand what this means, Suppose that you have a dataset D for a binary classification problem. Imagine that you have 50% of the data in class 0 and 50% in class 1. After the split, you want your D_T to have the same proportions between the classes. We will see in Chapter 8 what could go wrong if you train a model on a dataset with a very skewed distribution between the classes. Here are three tips to avoid the most common mistakes.

- If you have a classification problem, always compare the distributions of the classes in the original dataset D and in the D_T and D_V datasets. They should be similar.
- Always (apart from very specific cases) shuffle the data before doing any split.
- Read Chapter 8 for a deeper understanding of what could happen when training a model with a very unbalanced dataset.

In general **stratified sampling** is a method of sampling from a population (in our example, the original dataset D) which is partitioned into sub-populations (in a classification problem the sub-populations can be the classes) [1] (check Chapter 6 for more information on the topic). In general, stratification is the process of dividing members of the population into homogeneous subgroups before sampling[5].

Info Sampling in the Context of Machine Learning

Sampling in the context of machine learning refers to the process of selecting a subset of data from a larger dataset. This subset is then used to train, test, or validate a machine learning model. In most cases, sampling occurs by randomly selecting data points, ensuring that each data point has the same chance of being chosen. This method is simple but effective in creating a representative sample of the entire dataset.

Stratified sampling is used to ensure that various subgroups within a dataset are adequately represented in the sample. In this approach, the dataset is first divided into different *strata* (thus the name stratified) or subgroups based on a specific characteristic or criteria, such as age groups, income levels or other relevant factors. Once the dataset is divided into these strata, a random sample is taken from each subgroup. The size of the sample of each stratum can either be proportional to the size of the stratum in the entire dataset (*proportional stratified sampling*) or equal regardless of the stratum size (*equal stratified sampling*). By doing this, stratified sampling ensures that each subgroup is adequately represented in the final sample, reducing sampling bias, and improving the generalisability of the model's results. This technique is particularly useful in scenarios where certain subgroups may be under-represented in a random sample, leading to biased or skewed results. All this is discussed at length in Chapter 6.

As an example, let us suppose that we are trying to predict political preferences. Let us assume that in our dataset D we have people that can be clearly divided into three groups: very old people (let us call this group A), young people (B) and middle-

[5] Definition taken from `https://en.wikipedia.org/wiki/Stratified_sampling`.

aged (C). But suppose that in D we have different numbers of elements (which we will indicate with n_A for group A, n_B for B and n_C for C) in the three groups. How should we divide D into subgroups? For example, we could have $n_A = 1000$, $n_B = 10000$, and $n_C = 500$. To select 80% of the data, we could simply do a random sampling. By doing so, we run the risk of getting a sample that does not reflect the original population. But, for example, if we knew that in the country we actually have 30% of old people, 50% of middle-aged people and 20% of young people, we may want to sample from each subgroup according to these proportions to better reflect what could happen in the country, or, in other words, to make the sample (D_T) more representative not of the original dataset D, but of the country. So splitting a dataset is not as trivial as one may think, and it depends strongly on the research questions we are trying to analyse.

Info Sampling and Splitting in Python

In general, stratified sampling is very easy to implement in Python. For example, in `scikit-learn` it is as easy as using `sklearn.model_selection.StratifiedShuffleSplit`[a]. Splitting the dataset into two parts is even easier. This can be done with `train_test_split`[b]. Performing the necessary splitting can typically be done in one line of code. Checking the official documentation is always a good start.

[a] You can check the official documentation at `https://scikit-learn.org/stable/modules/generated/sklearn.model_selection.StratifiedShuffleSplit.html` (last accessed on 17 October 2023).

[b] You can check the original documentation at `https://scikit-learn.org/stable/modules/generated/sklearn.model_selection.train_test_split.html` (last accessed on 17 October 2023).

7.9 Monte Carlo Cross-Validation

If you try the hold-out approach a couple of times, you will soon realise that you get different results simply by splitting the datasets differently. Imagine splitting the data set into two parts N times. Let us indicate the training and validation parts as $D_T^{(i)}$ and $D_V^{(i)}$ with $i = 1, ..., N$. Let us also indicate a given metric m_i evaluated on $D_T^{(i)}$ for a model trained on the ith split ($D_T^{(i)}$). m_i could be the MSE or the accuracy, for example. It is easy to verify that m_i will be different from each other. This warrants the question: Which m_i is correct? Which should you choose to assess your model performance? The correct answer is that **all** are correct. An even better answer is that **it is not possible to properly assess the performance of the model with a single split**.

Warning Limitations of Assessing Model Performance with the Hold-Out Approach

Blogs, websites and tutorials (and many books and papers) try to convince you that to assess a model performance, the hold-out approach is good enough. You should not believe everything you read on the Internet. Using the hold-out approach may give you a superficial indication of the generalisation properties of a model, but it is a far cry from a serious assessment.
This problem commonly manifests itself even in papers published in prestigious scientific journals. Very often researchers want to show that their model is better than existing ones (a process called model selection or comparison, discussed at length later in the chapter). To do this, many will simply use the hold-out approach and compare the results with other models. If you compare, for example, the accuracy of two models that are (numbers are invented) 95.6% and 96.0% (your model is the one obtaining 96.0%) you cannot say that yours is better unless you properly study what happens if you split your dataset differently. Suppose that by performing the dataset split differently, your model will get an accuracy of 95.2%. What now? Is your model better or worse than the other? Model comparison is a tricky business, and we will discuss this later in this chapter. For now, it is enough to remember to be skeptical of results given with the hold-out approach. In the interest of fairness, the hold-out approach is **not** wrong. It simply gives only a very superficial assessment of a model performance that can be very misleading.
A wonderful alternative example of this problem (although not due to the dataset split) can be found in the paper by David Picard [2]. Although he studies the effect of random seeds on the performance of models in computer vision, the problem is the same. Picard shows that it is surprisingly easy to find models that perform much better or much worse than the average by simply changing the random seed that is used during the training of models, and thus one should be very careful when claiming exceptional performances or when claiming anything at all really.

You may think that if one single split is not enough, then with the Monte Carlo cross validation approach, we have a solution. You can simply calculate your metric for all the splits (thus getting multiple values) and give the standard deviation of all the values to give an idea of how much your metric varies if you split the data in different ways. To assess the performance of your model, you would give $\mathbb{E}(m_i)$ and $\sigma(m_i)$ over all the splits that in this notation are indicated with i. But this is also fraught with peril, as the next section will reveal. You should imagine where this is heading if you recall our discussion about the central limit theorem we did in Section 5.16. Let us discuss this a bit more.

7.9.1 How to use the Monte Carlo Cross-Validation Approach

To see what problems you may encounter, consider the following practical example. Consider a dataset $\{x_i, y_i\}_{i=1}^N$ with $x_i = 0.001 \cdot i$ and $y_i = x_i^2$ and N some integer greater than zero. We will suppose that we do not know how the dataset has been generated, so to start with, we will fit the data with a linear function (we will later use a polynomial model and compare it to the linear one). Consider the algorithm 1. If we run it, we will have at disposal N values of the MSE and we can study its

---

Algorithm 1 Monte Carlo Cross-Validation approach Algorithm Study

---

set N to some value
$j \leftarrow 1$
p is the percentage of data used for validation.
while $j < N$ **do**
  Split the dataset in two parts: $D_T^{[j]}$ and $D_V^{[j]}$ in proportions $1-p$ and p for training and validation respectively
  Train a linear regression model $M^{[j]}$ on $D_T^{[j]}$
  Evalute the MSE on $D_V^{[j]}$ and save its value in $\text{MSE}^{[j]}$
  $j \leftarrow j+1$
end while

---

distribution and properties by varying p and N. The reason we are doing this will become clear very soon. First, let us consider a large number N of data points. What does the distribution of $\text{MSE}^{[j]}$ look like? In Chapter 5 we discussed the central limit theorem that answers exactly this question. Let us again discuss this case to make it more clear.

Consider $N = 5 \cdot 10^4$ and $p = 0.3$. You can see the distribution of the MSE obtained with Algorithm 1 in Figure 7.7. What can we see? Well it is strangely similar to a normal distribution, and even if at this point you do not know why (see the central limit theorem in Section 5.16.2 to know why), for $N \to \infty$ the distribution will tend to a normal one. When N is small (we will never have an infinite number of data points, of course) we can still consider this distribution well approximated to a Gaussian (for all practical reasons). Already with N of the order of 10^2 the approximation starts to be reasonable. So far so good, but what happens when we change p? In Figure 7.8 you can see the results for two values of p: 0.3 (left panel) and 0.5 (right panel). They both resemble (in an approximate way) normal distributions but for $p = 0.5$ the distributions are narrower than for $p = 0.3$, in other words, their standard deviations are different. for $p = 0.3$ we get 0.0082 and for $p = 0.5$ 0.0053. But what is happening? Does the MSE distribution depend on the size of the validation dataset? Recall that p indicates the fraction of data that you are using to validate your model. It turns out that it does. This is due to the central limit theorem. If you are more mathematically inclined, you can check Section 5.16.2 for more precise explanations, but if you just want to get the gist of it, you have to remember that when using multiple splits and evaluating

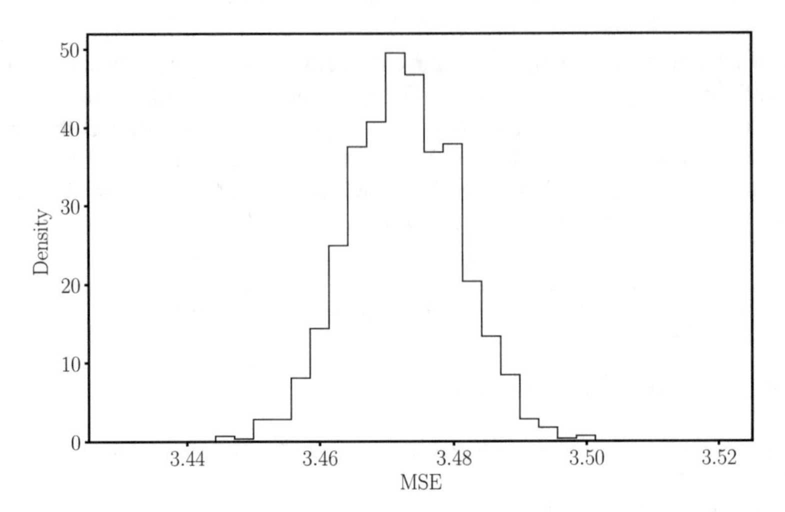

Fig. 7.7 The MSE distribution obtained with Algorithm 1 with $N = 5 \cdot 10^4$ and $p = 0.3$.

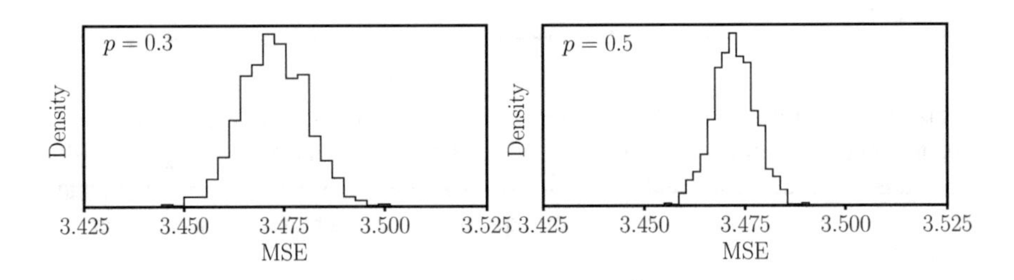

Fig. 7.8 The MSE distribution obtained with Algorithm 1 with $N = 5 \cdot 10^4$ and $p = 0.3$ (left panel) and $p = 0.5$ (right panel).

the MSE or accuracy in your validation data set, the standard deviations of their values will depend on the size of the validation dataset, and more precisely it will be proportional to the inverse square root of the number of data points in D_V. So, you should be very careful when comparing model performances by using averages and standard deviations. You can make the standard deviations smaller or larger simply by changing the proportions between your training and validation dataset. Why is this relevant? When you have a certain variable X and tell someone that you have its average $\mathbb{E}(X)$ and its standard deviation $\sigma(X)$, people will interpret that most of the data will be between $\mathbb{E}(X) \pm 2\sigma(X)$. Recall that assuming that X follows a Gaussian distribution, roughly 95% of its values are between $\mathbb{E}(X) \pm 2\sigma(X)$. Consider our example. And suppose that we use $p = 0.5$. When we say that with the Monte Carlo Cross-Validation approach we have $\mathbb{E}(\text{MSE}) = 3.472$ with $\sigma(\text{MSE}) = 0.005$, what we are insinuating (or better yet how it will be interpreted) is that a *typical* model (of the model type we choose, as adaboost, linear regression, or similar) will have as approximate highest value of the MSE roughly 3.477 **regardless of the data used!** But that is not correct. A much better formulation would be that *a typical model will have the highest MSE value of approximately 3.477* for a validation dataset of the size obtained for $p = 0.5$.

On the other hand, the average value is a reliable estimator of the expected value of the MSE that **does not** depend on the size of the validation dataset, but the standard deviation is very tricky to use and interpret properly. You can freely use the average of your metric (that does not depend on the size of the validation dataset), but you should be **very** careful when using the standard deviation.

Warning **Giving Results with the Monte Carlo Cross Validation Approach**

One way of assessing the performance of a model is to follow Algorithm 1. You can then give the average of your metric and its standard deviation. This is done often in scientific papers, but it makes interpretation of the results very difficult, because of the problem outlined above. Another aspect that we have not quite discussed in the previous sections is that in Algorithm 1 you train a new model for every new split. This means that you are not assessing the performance of **one single model**, but assessing the **potential** performance of a generic model type when trained on a data split in a given proportion. Let us rewrite this sentence in boldface again.

With the Monte Carlo Cross-Validation approach, you are not evaluating the performance of a single model, but assessing the potential performance of a generic model type (for example, linear regression) when assessed on a validation dataset of a given size.

This warrants the question of how to choose the **best** model then, and that is what is called **model selection** that will be discussed later in the chapter. Note that another downside of the Monte Carlo approach is that you must train a model many times. This can be very computationally intensive. If your model takes one day to train, you surely cannot consider 1000 splits. That is 2.7 years of continuous training. To address this problem, you can use the k-fold cross-validation approach described in the next section.

7.10 Monte-Carlo Cross Validation with Bootstrapping

In the previous section, we have described Monte-Carlo cross validation. You can use the same approach by using bootstrapping. The following process describes it.

1. You split your dataset D in two parts: D_T for training (say 80% of D) and D_V for validation (say 20% of D).
2. Create a large number N of bootstrap samples by sampling with replacement (see Section 6.8 for a discussion on bootstrap) from D_T.
3. Train N models, each on a different bootstrapped sample.
4. Evaluate the performance of N models on D_V.

With this process, you will have N estimation of the performance of the model that you can use to calculate its average.

I wanted to describe this method here, but I would not suggest using it. The main reason is that you do the initial split only once, therefore, there is a high probability for selection bias in the bootstrap samples, especially if the D_T is not perfectly representative of the population or is small. This can skew performance metrics. Additionally, there may be assumptions about the data distribution that do not hold in the bootstrapped samples, affecting the validity of the model evaluation. For example, if the original data have imbalances, these might be exaggerated or minimised in the bootstrapped samples in unpredictable ways. Do not use this approach and use more classical approaches such as k-fold or Monte Carlo cross-validation without bootstrapping.

7.11 k-Fold Cross Validation

k-fold cross validation is probably the most used, by far, of all cross-validation approaches. To use it, you first divide the dataset D in k portions D_i with $i = 1, ..., k$ (usually all the same size) so that $D_1 \cup D_2 \cup \cdots \cup D_k = D$ and $D_i \cap D_j = \emptyset$ for $i \neq j$ and $i, j = 1, \cdots, k$. Now, you loop over all D_i. You start with D_1 and use it as a hold-out set. You train your model on $D_2 \cup D_3 \cup \cdots \cup D_k$ and validate it on D_1. You will get one value of your chosen metric (we suppose here that you are tracking only one metric, for example, the MSE) which we will indicate here as m_1 (evaluated on the hold-out set D_1). Then you use D_2 as a hold-out set and you train your model on the union of all the remaining D_i for $i \neq 2$. You will then evaluate your model on D_2 and obtain an m_2 (evaluated on D_2). You continue in this fashion for all D_i. You will end up with k values for your metric. In Figure 7.9 you can see a diagram of the method. This approach is often used, especially if the training is computationally expensive. Typical values of k are 5 or 10.

Warning k-Fold Cross Validation Limitations and Assumptions

In general, there are some limitations in using this (and the hold-out approach, too) cross validation method. First, it assumes that the data points in the dataset are independent and identically distributed (i.i.d.). This assumption is crucial, as violation, such as in time series data where data points are

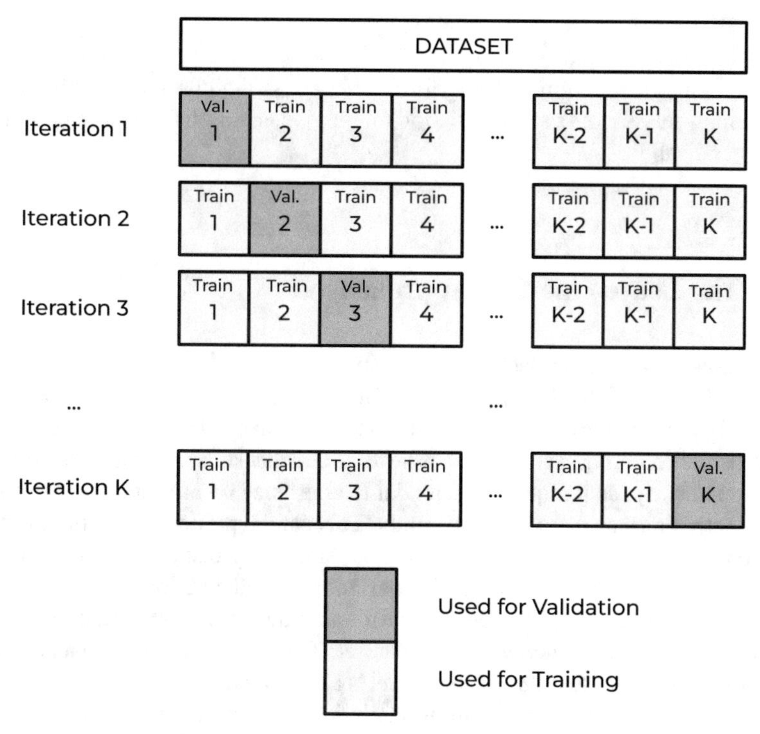

Fig. 7.9 k-fold cross validation approach divide a dataset D in k non-overlapping portions. By looping over all D_i and using each D_i as a hold-out set and training on the remaining portions merged, one obtained k values of any given metric that can be used to assess the generalisation property of a model type on a given datatype. In the figure, each portion of the dataset D_i is indicated by a box. Dark orange indicates the portion used for validation, while yellow indicate the portions used for training.

correlated, can lead to biased estimates of model performance. Secondly, the technique assumes that the data are representative of the underlying population and can be randomly sampled. This random sampling is crucial for the division of the data into k-folds to ensure that each fold is a good representative of the entire data set. If the data are not randomly sampled, the cross-validation results might not accurately reflect the model's performance on unseen data. Additionally, the distribution of the target variable and the explanatory variables (the features) should be consistent across all folds. Significant variation in the distribution between different folds can result in a model being trained and validated on unrepresentative subsets of data, leading to skewed performance metrics. Furthermore, there should be enough data to meaningfully split it into k folds. With very small datasets, dividing the data into too many folds can leave too little data for training, leading to high variance in the estimation of the model performance. Finally, the model being evaluated should not be overly sensitive to small changes in the training set. If a model's performance varies significantly with different training sets, k-fold cross-validation might give an unreliable estimate of its generalisability.

7.12 The Leave-One-Out Approach

The Leave-One-Out Approach (LOO) is simply a k-fold cross validation for $k = N$ (with N the size of the dataset). This is almost only done when the dataset is very small. I have seen it used in cases where the dataset has only 20-30 elements. LOO, while useful in certain scenarios, has several drawbacks. Primarily, it is computationally intensive, as it requires the model to be trained on almost the entire dataset for as many times as there are data points. This can be particularly burdensome for large datasets or complex models. Moreover, since only one observation is left out in each iteration, LOO can lead to high variance in model performance estimates.

The results can be overly optimistic for small datasets, as the training set used in each fold closely resembles the full dataset. This closeness can also lead to poor generalisation error estimates, as the model is almost always trained on a dataset very similar to the one it is tested on. Additionally, LOO is not suitable for datasets where observations are not independent, such as time-series data. Another concern is the sensitivity of LOO to outliers. Since each data point is used as a test set, outliers can have a disproportionate impact on the model's evaluated performance. This can skew the results, leading to either an overestimation or an underestimation of the model's capabilities. Due to these challenges, other cross-validation techniques, such as k-fold or stratified cross-validation, are often preferred, especially in the cases of large or complex datasets.

7.13 Choosing the Cross-Validation Approach

Choosing the appropriate cross-validation method in machine learning is crucial for estimating generalisation characteristics of models (or model types). Here are some guidelines and tips.

k-Fold Cross-Validation:

- *When to use:* k-fold is suitable for datasets that are large enough to be split into multiple smaller sets, ensuring each fold can serve as a meaningful test set. It works well with fairly balanced datasets, which means that each class or outcome of interest is roughly equally represented across the dataset. When the goal is to obtain a more accurate estimate of model performance than what a simple hold-out approach would provide, k-fold cross validation helps in

averaging the performance across multiple subsets. Additionally, when comparing different models or different configurations of the same model, k-fold cross validation provides a more reliable basis for comparison than a single train-test split (more on that later).
- *Tip:* choose k carefully; generally, $k = 10$ is a good balance. Remember to apply the same pre-processing steps to each fold as you would to your entire dataset. This consistency is vital to avoid introducing biases or variances that could skew your model's performance evaluation (remember our discussion about data leakage?).

Stratified k-Fold Cross Validation:

- *When to use:* when dealing with datasets where some classes are underrepresented, stratified k fold ensures that each fold maintains the same proportion of each class as in the full dataset. This is crucial to maintaining the integrity of the model's performance across classes. It is most beneficial in classification problems where maintaining the distribution of classes is important. For example, in medical diagnosis datasets where one outcome is much rarer than another, stratified k-fold helps in preserving this imbalance across all folds. Because of the stratification, it offers a more accurate reflection of the model's ability to handle class imbalances. When tuning hyper-parameters (more on this later), especially in imbalanced classification problems, a stratified k-fold provides a more consistent and fair evaluation of each set of parameters.
- *Tips:* apply the same pre-processing steps to each fold as you would to your entire dataset. This consistency is vital in avoiding introducing biases or variances that could skew your model's performance evaluation. In datasets with very small minority classes, be cautious. Stratified k-fold cross validation can still be used, but it is important to ensure that each fold gets a representative sample of the minority class, which can be challenging if the class is extremely small. If reproducibility is important, set a random state for the splitting process. This ensures that you get the same splits every time you run your code, which is crucial for consistent and comparable model evaluation.

Leave-One-Out Cross-Validation (LOOCV):

- *When to use:* LOOCV is particularly useful when you have a very small dataset. Since it maximises the amount of training data by only leaving one sample out at a time, it is beneficial when each data point's inclusion in training significantly impacts the model's learning. LOOCV can be a good choice if the dataset is balanced and the samples are relatively homogeneous, which means that the exclusion of a single sample does not significantly change the nature of the training set.
- *Tips:* Can be very time consuming with larger datasets. Although the evaluation variance is low in terms of training data, LOOCV can still result in high variance in model performance if the model is very sensitive to the specific data points used for training (for example, in the case of extreme outliers).

Because nearly all data are used for training, there is a higher risk of overfitting, especially with flexible models.

Monte Carlo Cross-Validation (with or without boostrap):

- *When to use:* it is suitable for large datasets. Monte Carlo cross-validation allows for a quicker assessment of the model performance. Unlike k-fold cross-validation, random subsampling allows flexibility in choosing the proportion of the dataset to be used as training and test sets. This can be particularly useful when working with very large datasets (for example, if you have enough data you could use 99% of the data for training and just 1% for validation, if 1% contains enough data and is representative of the entire dataset you are good to go). It offers a good balance between computational efficiency and the stability of the performance estimate. The number of repetitions can be adjusted based on the available computational resources and the desired confidence in the model's evaluation.
- *Tips:* the number of repetitions should be high enough to ensure stability in the performance estimates. A common practice is to use at least 30-50 repetitions. If reproducibility is important in your analysis, ensure that you set a random seed for each subsampling iteration. When dealing with unbalanced data sets, it is important to ensure that each random subsample maintains the class distribution. Techniques like stratification can be applied within each subsample.

This section should give you some guidelines and tips that should help in choosing the right cross-validation approach. In the next section, we will discuss how to compare models and optimise their parameters.

7.14 Model Selection

Now let us turn our attention to discussing how to compare models. In general, in machine learning, one is faced with a huge variety of models. For example, for a classification problem, you may choose logistic regression, support vector machines, decision trees, random forest, and many more. In addition, each model has parameters that can be changed and optimised. How to choose the best model or the best parameters in a given model class? This question pertains to the **model selection** problem. We will also mainly consider in this discussion supervised learning and will only briefly touch upon unsupervised learning model selection later at the end of the chapter.

Before discussing model selection, let us summarise (you will find a long discussion on this topic in Chapter 9, but to make this chapter as standalone as possible, we will repeat some of the information here) the difference between the terms **parameter** and **hyper-parameter**.

Definition 7.4 (Hyper-Parameters) Parameters that are related to the model (for example, depth in decision trees, or the learning rate) and are not influenced by the training process (thus, do not depend on the training data) are called **hyper-parameters**.

Parameters can be defined as follows.

Definition 7.5 (Parameters) **Parameters** indicates what is changed during the training process (for example, the weights of a neural network or the slope in a linear regression problem).

7.14.1 Model Selection with Supervised Learning

Now let us suppose that we have q models $\mathcal{M}_i$ with $i = 1, \cdots, q$ and that we are trying to solve a supervised learning problem. Two models may be different simply because we are considering a different set of hyper-parameters. For example, we may consider two neural networks, one with one hidden layer and one with two. We will now discuss how to apply the hold-out approach to this problem. But we will need **three** datasets, not just two. The reasons will become clear very soon. Typically, size ratios are 60%/20%/20% or 70%/15%/15%. We split our dataset D in three parts: D_T for training, D_V for validation, and D_{test} for testing. To select the best model, the following process should be followed.

1. Train all q models $\mathcal{M}_i$ on D_T. We will assume that we want to track a single metric $m_i(\cdot)$, as, for example, the MSE (the subscript i indicate that the metric is being evaluted from the model $\mathcal{M}_i$. Its value on a given dataset D will be indicated with $m_i(D)$.
2. Compute the metric on the validation dataset $m_i(D_V)$.
3. Select the model j that has the lowest value for the metric $m_i(D_V)$. Note that if you are considering the MSE for a regression problem, you are looking at the lowest value, if you are considering a classification problem and you are considering the accuracy, you want the highest value of the metric.
4. Compute the metric on the test dataset $m_i(D_{\text{test}})$.
5. Compare $m_i(D_{\text{test}})$ with $m_i(D_V)$. If the two values are similar, you are good and you can finally train your chosen models on the entire dataset, ready to be used on real data. If the two metrics are different, then it may happen that your hyper-parameters are perfect for your test dataset but not for unseen data. Basically, it means that you overfitted the test dataset with your hyper-parameters. In this case, you need (as we discussed) to consider a less complex model and start again.

Warning **Number of Datasets**

You will need to split your dataset in three portions **only** if you are doing hyper-parameter tuning. If you simply want to validate a class of models, then two portions will be the number you need. Remember, you may overfit a model to a dataset not only by training, but also by choosing values for the hyper-parameters that works well **particularly** for the given test dataset but would not lead to a well generalising model. This is the reason why you need three portions of a dataset when you are performing hyper-parameter tuning.

The steps described above refer to generalising the hold-out approach to model selection. If you want to use k-fold cross validation you need to slightly modify the process. In this case, you will need to first split the dataset in two portions: training and test. Then you apply the k-fold cross validation to your training dataset. The metric $m_i(D_V)$ that you will need to compare the performance of models will be in this case the expected values of the metric over all folds. The rest of the approach works exactly the same as described above.

7.14.2 ■ Model Selection with Unsupervised Learning

Model selection in unsupervised learning, where there are no labelled outcomes, involves different considerations compared to supervised learning. In the list below you will find few examples, but we will not discuss it in general in this book, since it is strongly model specific[6]. We will only explain some key ideas for clustering since it is widely used in different scientific fields.

Note that this is not, by any stretch of the imagination, an exhaustive list, and has the only goal of giving some information for the reader to deepen her or his knowledge on the topic.

1. **Visual Inspection**: Techniques like t-SNE or PCA are used for visual inspection of clustering results, providing insights into data grouping and separation. The one that visually group the data better, can be chosen as the best model.
2. **Consensus and Stability**: Assessing the stability or consensus of the model across different runs, initializations, or data subsets can be used for model selection. For example, depending on your requirement, a model that is more stable on multiple datasets or conditions may be chosen over one that has a better performance but is not as stable.
3. **Domain Knowledge and Interpretability**: Ensuring that results are interpretable and meaningful within the application domain is a key factor in model selection. Models that are exaplainable (and not black boxes as neural networks) may be preferred to models that cannot be explained.

[6] Remember that the goal of this book is to explain fundamental concepts that are generally valid and not model specific.

Info ■ Model Selection in Clustering

Model selection in clustering involves two tasks: choosing the most appropriate algorithm and determining the optimal number of clusters for a dataset. Here is a list of the most important approaches (and some references for the interested reader).

1. **Determining the Number of Clusters**: Methods like the Elbow Method, the Silhouette Coefficient, and the Gap Statistic help estimate the optimal number of clusters [3, 4].
2. **Choosing the Right Algorithm**: The choice of algorithm depends on the dataset characteristics. K-Means is suitable for spherical clusters, while DBSCAN works well for varied densities [5, 6] for example.
3. **Evaluating Cluster Quality**: Internal indices such as cohesion and separation are used to assess cluster quality [7].
4. **Scalability and Efficiency**: The computational efficiency and scalability are important considerations, especially for large datasets [8].
5. **Data Characteristics**: The distribution, scale, and dimensionality of the data influence the choice of the clustering model [9].
6. **Domain Knowledge**: Incorporating domain knowledge can guide the model selection and validation process [10].

7.15 ■ Qualitative Criteria for Model Selection

While we have focused primarily on quantitative approaches for selecting one model over others, it is crucial to recognize that, in many instances, qualitative criteria are just as important, if not more so, than quantitative ones. The following list outlines key qualitative criteria that play an important role in model selection.

1. **Interpretability and Explainability**: The ability of stakeholders to understand and trust model predictions is crucial. For example, in medical diagnostics, clinicians prefer models like decision trees that provide transparent reasoning over black-box models like deep neural networks.
2. **Domain Suitability**: The model's alignment with specific domain knowledge and its practical implications are vital. In finance, for instance, models must adhere to regulatory and compliance norms, which often require domain-specific tailoring.
3. **User Trust and Acceptance**: User confidence in a model's outputs can be a deciding factor, particularly in consumer-facing applications. A simpler model with lesser accuracy might be favored for its ease of use and understanding.
4. **Robustness and Sensitivity**: Assessing how a model responds to variations in data is key. In autonomous driving, models must be robust to changes in weather conditions, lighting, and unexpected road scenarios for example. So sometimes a

more stable model to changes in the data is to be preferred to one that has a better performance but is less stable.

5. **Ethical and Social Considerations**: Models must be evaluated for fairness and bias. In recruitment tools, algorithms must be scrutinized to prevent gender or racial bias in candidate selection.
6. **Scalability and Flexibility**: The ability of a model to adapt to changing data volumes and patterns is essential. In e-commerce, recommendation systems must scale with increasing user base and rapidly changing consumer preferences.

Qualitative criteria are indispensable in model selection, ensuring that models are not only technically sound but also ethically responsible, understandable, and practical.

References

1. Changbao Wu and Mary E Thompson. *Sampling theory and practice*. Springer, 2020.
2. David Picard. Torch.manual_seed(3407) is all you need: On the influence of random seeds in deep learning architectures for computer vision, May 2023. arXiv:2109.08203 [cs] version: 2.
3. Peter J Rousseeuw. Silhouettes: a graphical aid to the interpretation and validation of cluster analysis. *Journal of computational and applied mathematics*, 20:53–65, 1987.
4. Robert Tibshirani, Guenther Walther, and Trevor Hastie. Estimating the number of clusters in a data set via the gap statistic. *Journal of the Royal Statistical Society: Series B (Statistical Methodology)*, 63(2):411–423, 2001.
5. J MacQueen. Some methods for classification and analysis of multivariate observations. *Proceedings of the fifth Berkeley symposium on mathematical statistics and probability*, 1(14):281–297, 1967.
6. Martin Ester, Hans-Peter Kriegel, Jörg Sander, and Xiaowei Xu. A density-based algorithm for discovering clusters in large spatial databases with noise. In *Kdd*, volume 96, pages 226–231, 1996.
7. Leonard Kaufman and Peter J Rousseeuw. *Finding groups in data: an introduction to cluster analysis*. John Wiley & Sons, 2009.
8. Anil K Jain, M Narasimha Murty, and Patrick J Flynn. Data clustering: A review. *ACM computing surveys (CSUR)*, 31(3):264–323, 1999.
9. Charu C Aggarwal and Chandan K Reddy. *Data clustering: algorithms and applications*. CRC Press, 2013.
10. Jiawei Han, Micheline Kamber, and Jian Pei. *Data mining: concepts and techniques*. Elsevier, 2011.

Chapter 8
Unbalanced Datasets and Machine Learning Metrics

One out of four people in this country is mentally unbalanced. Think of your three closest friends; if they seem OK, then you're the one.

Ann Landers

This chapter explores the challenges and solutions associated with unbalanced datasets in machine learning. It begins by defining what constitutes an unbalanced dataset and emphasises their prevalence. The chapter introduces the concept of machine learning metrics, vital for evaluating the performance of models trained on such datasets. It presents a simple example to illustrate how traditional models fail in the face of extreme class imbalances and how this leads to misleading accuracy metrics. The core of this chapter delves into various approaches to address dataset imbalance, focusing primarily on data level approaches like oversampling and undersampling. It discusses the advantages and disadvantages of these techniques and highlights their practical applications through examples. The Synthetic Minority Oversampling Technique (SMOTE) is introduced as a sophisticated method to generate synthetic samples to balance datasets. Moreover, the chapter covers crucial metrics for assessing model performance in the context of unbalanced datasets, including the confusion matrix, sensitivity, specificity, precision, F_β-score, and balanced accuracy. The discussion extends with the Receiving Operating Characteristic (ROC) curve and the Area Under the Curve (AUC), providing a comprehensive framework for evaluating and enhancing model performance in situations of class imbalance.

8.1 Introduction

Consider a binary classification problem where each observation can be classified into one of two distinct classes. These classes can be represented in various ways: as 0 and 1, as 'dog' and 'cat', or as 'high' and 'low', among others. In this chapter, we will represent these classes numerically as 0 and 1. Depending on your specific problem, there will be a certain number of observations in class 0, denoted as m_0, and a certain number in class 1, denoted as m_1. Based on this, we provide the following general definition:

U. Michelucci, *Fundamental Mathematical Concepts for Machine Learning in Science*,
https://doi.org/10.1007/978-3-031-56431-4_8

Definition 8.1 A dataset is said to be **unbalanced** if the target variable has more observations in one class than in the others. Or, in other words, when $m_0 \neq m_1$.

It is important to explain why it is essential to discuss this. First, you will almost always have to deal with *unbalanced* datasets. Some will be *more* and some *less* unbalanced (an intuitive understanding of what more and less means will suffice for now). Especially in medicine, this is always the case. If you are trying to predict a disease, or to predict onset of medical conditions, you will always have (and we all hope this is always the case) more healthy patients than sick ones. A concrete example can be found in the article by Cohen *et al.* [1]. In the paper they consider a dataset with 75 patients with infections they acquired in the hospital (let us assign to this group class 0), and 608 without (class 1). Class 0 contains only 11% of all patients, while class 1 contains 89%. This is a perfect example of a *unbalanced dataset*. This paper has many good examples of all the things that could go wrong (the authors discuss those problems explicitly). I like some tension, so I will not immediately reveal what could go wrong. Let us illustrate this with a simple example.

8.2 A Simple Example

We will build a dataset for our testing and see what can go wrong. Each input observation $\mathbf{x}_i$ will have two features (this will make visualisations easier).

$$\mathbf{x}_i = (x_{1,i}, x_{2,i}). \tag{8.1}$$

Points in class 0 will be generated by sampling $x_{1,i}$ and $x_{2,i}$ from $\mathcal{N}(1, 1)$[1]. Points in class 1 by sampling the components from $\mathcal{N}(3, 1)$. Suppose now that we want to build a model to classify the inputs $\mathbf{x_i}$. We will use a linear support vector classifier (SVC). If you do not know what this is, don't worry. The models we are using are irrelevant. It is enough to know that a linear SVC will find a linear decision boundary in the $(x_{1,i}, x_{2,i})$ space. To better understand it, consider Figure 8.1. In the figure you can see 1000 points in class 0 and 1000 in class 1 (generated as described above). The diagonal line is what is called a decision boundary and has been obtained by training a linear SVC model. The background colours should be interpreted in the following way: the model will classify every point in the white region as class 1, while it will classify every point in the grey region as class 0.

The model works quite well. Most of the white points and the grey points are correctly classified. There are a certain number of points around the decision boundary that could not be classified correctly. We cannot do anything about it since we have points in classes 0 and 1 that overlap. However, this model reaches 91% accuracy. This is the behaviour that we would like to see. Now let us consider a new dataset: 1000 points in class 0 but only 10 in class 1. This is an extremely unbalanced dataset.

[1] Remember that with $\mathcal{N}(\mu, \sigma^2)$ we indicates a normal distribution with average μ and variance σ^2.

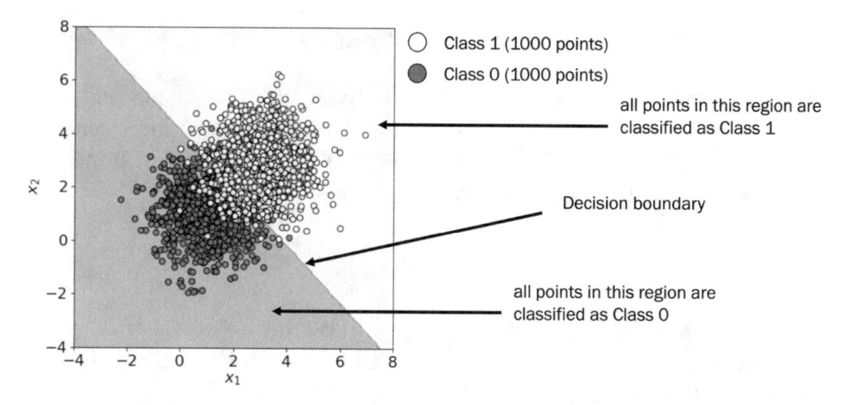

Fig. 8.1 You can see 1000 points in class 0 and 1000 in class 1 (points in class 0 have been generated by sampling $x_{1,i}$ from $\mathcal{N}(1, 1)$, points in class 1 by sampling $x_{2,i}$ from $\mathcal{N}(3, 1)$. The diagonal line is what is called a decision boundary and has been obtained by training a linear SVC model. The background colours should be interpreted in the following way: the model will classify every point in the white region as class 1, while it will classify every point in the gray region as class 0.

Let us again train the same model and check what decision boundary the model finds. You can see the results in Figure 8.2. You can see how the model simply ignores the 10 white points (class 1). It will classify all points as class 0. 10 points are not enough to really influence the learning phase of the model. This is the typical result when you try to train a model (neural networks, too) on an extremely unbalanced dataset. The model will simply learn to classify all inputs as being in the majority class.

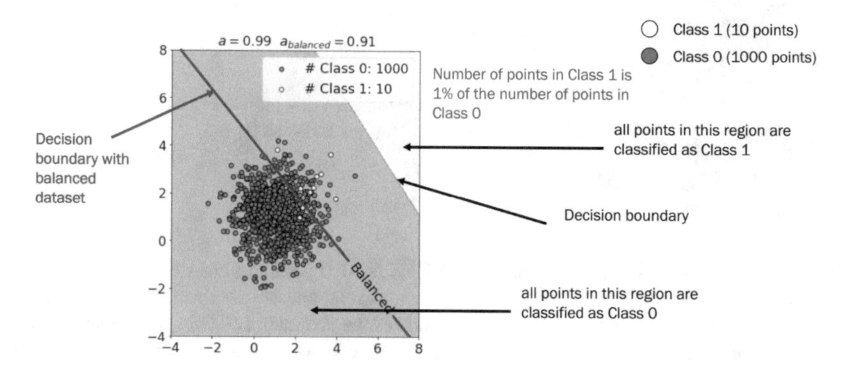

Fig. 8.2 In this case we trained again a Linear SVC with 1000 points in class 0 and only 10 in class 1. You can see how the decision boundary is now very far from the one we obtained with a perfectly balanced dataset (plotted in red). a indicates the accuracy, and a_{balanced} the accuracy we had when training the model with a balanced dataset of 1000 points in class 0 and 1000 in class 1.

Warning Accuracy can be misleading

Note that we identified the problem because we plotted the points and the decision boundary. Suppose that you simply check the value of the accuracy that the model can reach. Since your trained model will classify all inputs into class 0, you would get an accuracy equal to

$$a = \frac{1000}{1000 + 10} = 0.99 \tag{8.2}$$

so at first sight your model is doing really well! But if your goal is to identify those 10 points (they could be patients with a specific disease), then you are failing spectacularly.
When dealing with unbalanced datasets, you need techniques to check and assess how your model is really doing. This is what we will discuss in this chapter.

When increasing the number of points in class 1 to 50, the decision boundary that the model finds moves closer to the balanced one, as you can see in Figure 8.3. The

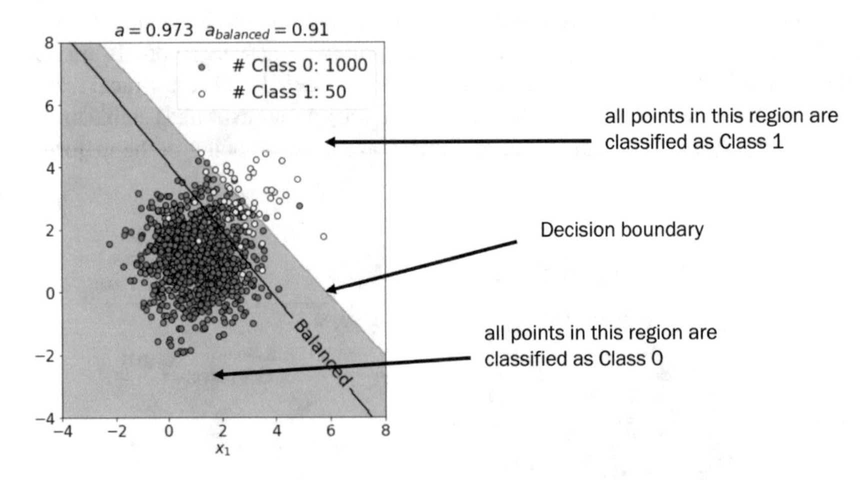

Fig. 8.3 In this case we trained again a Linear SVC with 1000 points in class 0 and 50 in class 1. You can see how the decision boundary is now closer to the one we obtained with a perfectly balanced dataset (compare it with Figure 8.2). a indicates the accuracy and a_{balanced} the accuracy we had when training the model with a balanced dataset of 1000 points in class 0 and 1000 in class 1.

result is not ideal, but it shows very clearly how things can go wrong. Note that this behaviour is more common than you might think.

Let us return to the article we discussed at the beginning about infections in hospitals. The authors published a table with results which is reproduced in Table 8.1.

Table 8.1 Baseline performance reproduced from [1]. The results show sensitivity, specificity and accuracy by training different models on the unbalanced dataset (original class distribution: 0.11 pos, 0.89 neg).

Classifier	Sensitivity	Specificity	CWA	Accuracy
IB1	0.19	0.96	0.38	0.88
Naive Bayes	0.57	0.88	0.65	0.85
C4.5	0.28	0.95	0.45	0.88
AdaBoost	0.45	0.95	0.58	0.90
SVM	0.43	0.92	0.55	0.86

The results show sensitivity, specificity and accuracy by training different models on the unbalanced dataset. If you check the accuracy column, you can see, for example, that an SVM (Support Vector Machine) model obtains an accuracy of 0.86, which is strangely close to 0.89 percent of patients in class 1 (healthy). When you see an accuracy very close to the percentage of inputs in the majority class, you should become suspicious. The model is not giving enough weight to the minority class, as we have seen in our toy example. Note that every model obtains an accuracy extremely close to 0.89, the percentage of patients in class 1. This problem affects all kinds of model and algorithms. Neural networks are not spared, of course.

We have not yet looked at sensitivity and specificity, but those are two metrics that will help understand what is going on. In this case, it is sufficient for you to know that sensitivity measures the ability to identify patients with an infection. For SVM, the value is only 0.43. That means that the model will identify only 43% of the patients with an infection, quite a bad result, don't you think? For IB1 (that is, a variation of the k-nearest neighbour algorithm) it is even worse, with a sensitivity of 0.19. A model trained with the unbalanced dataset is not able to find patients with infections (the main goal of the model in the first place). This behaviour is more or less extreme, depending on how unbalanced the dataset is. Remember our toy example? Extreme cases (only 1% of inputs in one class) will give models that basically classify everything in the majority class. The less unbalanced the dataset is, the less extreme this effect will be. However, you need to know how to check if something like this is going on and you need to have some techniques in your arsenal to deal with such situations.

If you think that unbalanced datasets are the exception, think again. They are the norm. Balanced datasets exist only in tutorials that you find on blogs and websites. Real datasets are **always** unbalanced in some measure.

Info Unbalanced Datasets: Classification and Regression

One of the first steps in any machine learning project is to understand the data you have at disposal. One of the most important questions is how (and not if) your data are unbalanced. Recall that we are only talking about a classification

problem. The more unbalanced your dataset is, the more your model will ignore the minority class. You can check this with many techniques that we will discuss in this chapter, but a first indication that this is happening is if your accuracy is equal (or very close) to the percentage of inputs in your majority class. If this happens, you should be suspicious and check what is going on.
Note that this problem does not appear only in a classification problem. In regression, it shows itself in a different and more subtle way. Suppose that you are trying to predict a continuous variable y. Imagine that your dataset contains 90% of points with y values in the interval $[0, 1]$ and only 10% in the interval $[1, 2]$. What will happen is that your model predictions will be bad in the interval $[1, 2]$, as it has not seen many data points in that interval. Although the problem is different in this case, it is something that you always have to check if you are doing regression. The techniques we will discuss in this chapter apply to a classification problem. However, we will also explore what actions can be performed in such situations toward the end of our discussion.

8.3 Approaches to Deal with Unbalanced Datasets

Several excellent reviews and articles have been published in recent years. For example, He and Garcia [2] published in 2009 a systematic review of metrics and algorithmic level approaches. Specifically, for classification. Sun *et al.* published a nice review in the same year [3]. In [4], García discusses at length the topic of how to process data for unbalanced datasets. Finally, Branco has compiled a large review [5] that covers many aspects. The reader has enough material here to learn enough about dealing with unbalanced datasets, in their most intricate details.

To deal with unbalanced datasets there are three main approaches.

- **Data Level Approaches** involve preprocessing the data in some way to restore class balance [6]. The idea behind them is to balance the dataset in some way. The two main methods in this category are *oversampling* (described in Subsection 8.3.1) and undersampling (described in Subsection 8.3.2). In this category, you can also find *feature selection* approaches that consists in processing the features to perform a kind of oversampling [7]. This approach is not yet much explored and goes beyond the scope of this book. But if interested, the reader is referred to the following publications for a few examples [7, 8, 9].
- **Algorithm Level Approaches** involve modifying (or designing) algorithms, so that they can handle unbalanced data. Generally, those methods can be categorised in two groups, one called cost-sensitive learning and the second ensemble learning. The first addresses the problem by modifying how the misclassification costs are calculated (for example, by modifying the function that is minimised in the

learning phase, in the case of neural networks the loss function[2]), in the second, multiple algorithms try to solve the problem at the same time, and then the result is obtained by some kind of vote between the different models.
- **Hybrid Approaches** consist in mixing the previous ones in various forms.

In this chapter (and book), we deal exclusively with data level approaches. There are several algorithms that deal with data, but they fall mainly into two categories: oversampling and undersampling (and variations). The interested reader can consult [11, 12, 13, 14, 15, 16, 17, 18] for a more in depth review of the main methods. In this chapter, we will discuss the main ideas on how to handle unbalanced datasets, but we will just scratch the surface. Readers interested in algorithm level approaches can consult [19, 20]. If you are interested in how to make deep learning work better with unbalanced datasets the following reference is interesting [21]. You may be interested, from a more practical point of view, in how to deal with unbalanced datasets with TensorFlow. If that is the case, you can consult the official documentation [10].

8.3.1 Oversampling

The first method you can use to deal with unbalanced dataset is called *oversampling*[3]. It simply consists of duplicating the inputs in the minority class enough times until your dataset is again balanced. In our toy example, when we had 1000 points in class 1 and 10 in class 0 we would simply consider the 10 points in class 0, 100 times. In Figure 8.4 you can see the results with our toy dataset when using oversampling. It is evident how oversampling really helps. Now the decision boundaries are very close to the on obtained with a balanced dataset. Furthermore, the accuracy reached with oversampling is now 89% with 10 points in class 1 and 92% with 50 points in class 1. Values that are very close to those we obtain with a balanced dataset.

Info Oversampling Advantages and Disadvantages

Advantages

- One has a larger dataset to train, and one does not ignore any input.

Disadvantages

- The model is worse in generalising, since it sees fewer inputs in class 1.

[2] One commonly used method when training neural networks with unbalanced datasets, is the use of weighted classes. Basically, the contributions in the loss function coming from the minority class will be multiplied by a factor that is inversely proportional to the class ratios. For a practical example, the reader is directed to the official Python TensorFlow documentation [10].

[3] Oversampling is easy to implement in Python with the library `imbalanced-learn`. A good overview can be found in the official documentation [22].

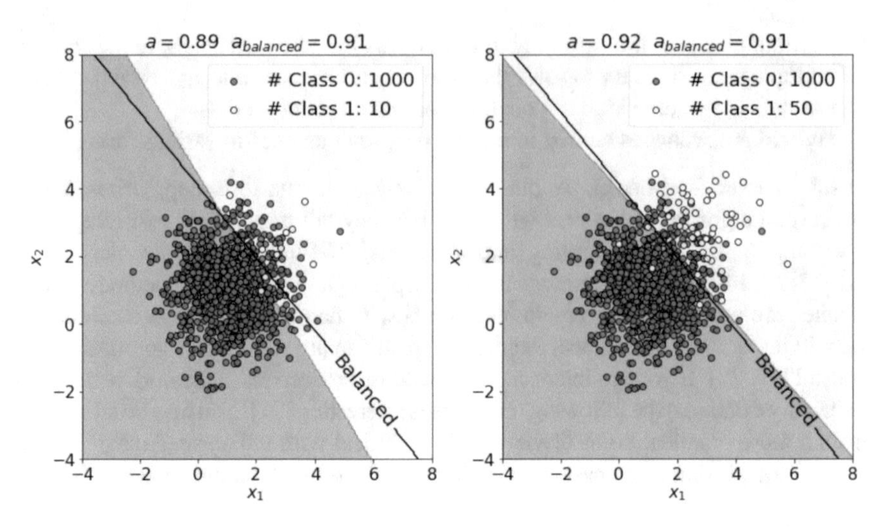

Fig. 8.4 Here are the results of our toy example with 1000 points in class 0 and 10 in class 1. Oversampling has been used to train the model. That means that the dataset used has the original 1000 points in class 0 and 100 times the 10 points in class 1.

In the previous example, we have simply replicated many times the data in the minority class, but there are smarter ways to get more observations from the minority class. An example of how to do that is by using what is called Synthetic Minority Oversampling TEchnique (SMOTE), proposed by Chawla *et al.* in 2002 [23]. SMOTE is a technique to address class imbalance in a dataset by generating synthetic samples of the minority class. It works by selecting a minority class sample and then generating synthetic samples along the line segments connecting it to its k-nearest minority class neighbours (see Section 8.4)

Example SMOTE in Python

Note that almost all the techniques explained in this chapter are available in Python in the library `imblearn` [24]. Given a dataset composed by numpy arrays `X_train` and `y_train` it is as easy as writing the following code (which we will not explain here):

```
from imblearn.over_sampling import SMOTE
sm = SMOTE(random_state=42)
X_res, y_res = sm.fit_resample(X_train, y_train)
```

Using the methods in Python is very easy, but without knowing how they work and when it makes sense to apply them, it is a futile exercise. The difficulty is not implementing the methods in Python but knowing when to use them and how to interpret the results.

There are many other methods that generate artificial observations from the minority class, by looking at the distribution of the features of the available data. These go beyond the scope of this book and therefore will not be discussed further.

8.3.2 (Random) Undersampling

The second method you can use to deal with an unbalanced dataset is called *random undersampling*[4]. It consists of randomly choosing as many inputs from the majority class as inputs in the minority class. In our toy example, when we had 1000 points in class 1 and 10 in class 0 we would simply consider the original 10 points in class 1, and we would choose randomly 10 points from class 0. In Figure 8.5 you can see the results with our toy dataset when using undersampling. It is evident

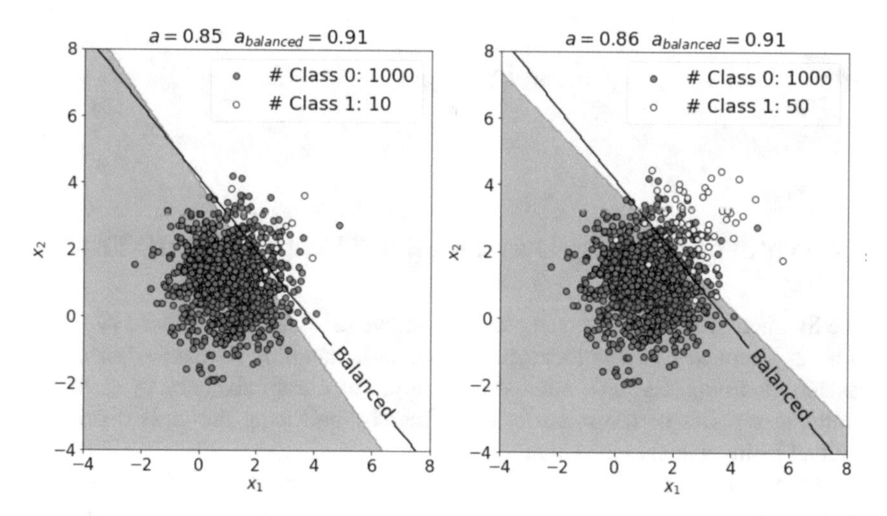

Fig. 8.5 Here are the results of our toy example with 1000 points in class 0 and 10 in class 1. Undersampling has been used to train the model. That means that the dataset used has the original 10 points in class 1 and 10 randomly chosen from class 0.

how undersampling really helps. The decision boundaries are now very close to the one obtained with a balanced dataset. Furthermore, the accuracy reached with oversampling is now 85% with 10 points in class 1 and 86% with 50 points in class 1. Values that are very close to those we obtain with a balanced dataset.

[4] Undersampling can be easily implemented in Python with the library `imbalanced-learn`. If you are interested, you can check the official documentation in [25].

Info Undersampling Advantages and Disadvantages

Advantages

- Data are not duplicated.

Disadvantages

- The dataset is much smaller, and one loses many inputs in the majority class.
- If the original dataset is small, this approach may be difficult to apply since one would lose too much data.

Warning Which Method should I use?

Note that there is no golden rule on how to decide which method to use. You must try both, check how they are doing, and decide which one works best for your case. This is a golden rule in machine learning: there are no fixed rules on how to choose parameters, algorithms, etc. You always must try and test what works for your case.

8.4 Synthetic Minority Oversampling TEchnique (SMOTE)

The Synthetic Minority Oversampling Technique (SMOTE), introduced by Chawla et al.[26], represents a significant advancement in handling unbalanced datasets in machine learning. SMOTE addresses the issue of class imbalance by generating synthetic samples of the minority class, thereby balancing the class distribution without losing valuable information.

The core idea of SMOTE is to create synthetic samples in the feature space of the minority class based on existing instances. This process involves the following steps:

1. For each sample in the minority class, identify its k nearest neighbours in the feature space.
2. Randomly select one of these neighbours and compute the vector difference between the feature vectors of the neighbour and the original sample.
3. Multiply this vector difference by a random number between 0 and 1, and add it to the feature vector of the original sample.
4. The resultant vector is a new synthetic sample that is added to the dataset.

In this way, one can create additional inputs from the minority class to reduce the imbalance in the dataset.

Info SMOTE Advantages and Disadvantages

Advantages

- **Mitigating Overfitting:** By generating synthetic samples, SMOTE helps in reducing the overfitting problem that often occurs with simple oversampling.
- **Enriching Feature Space:** SMOTE introduces more variety in feature space, making the decision boundary for classifiers smoother and more generalised.

Disadvantages

- **Noise Amplification:** If the minority class samples have noise, SMOTE might amplify this noise by generating synthetic samples based on these noisy instances.
- **Computational Complexity:** The process of finding nearest neighbours and generating new samples adds computational complexity to the data preprocessing step.
- **Not a Panacea:** While SMOTE is effective in many cases, it is not suitable for all datasets, especially those with highly complex or sparse minority class distributions.

SMOTE has been successfully applied in various domains, including but not limited to finance (for fraud detection), healthcare (for rare disease diagnosis), and social media (for sentiment analysis). The technique is versatile and can be adapted to most classification problems dealing with imbalanced datasets. In conclusion, SMOTE stands as a powerful and versatile technique to address class imbalance in machine learning. Its ability to generate synthetic, yet plausible, samples of the minority class makes it a valuable tool in the data scientist's arsenal, especially in scenarios where collecting more data is impractical or impossible.

8.5 ■ Summary of Methods for Dealing with Unbalanced Datasets

We have just scratched the surface of how to handle unbalanced datasets. Many methods and variations are available. To provide the reader with some more information, below is a list of some approaches and references where to find more information.

1. **Random Oversampling and Undersampling:** Simple yet effective techniques that involve replicating minority class instances (oversampling) or removing some from the majority class (undersampling) [27].
2. **Cluster-Based Oversampling:** An advanced oversampling method that involves creating synthetic samples in clusters of the minority class.[28]

3. **SMOTE and its Variations:** Synthetic Minority Oversampling Technique and its derivatives like Borderline-SMOTE and ADASYN [29, 30, 31].
4. **Cost-Sensitive Learning:** Adjusting the learning algorithms to make the misclassification of minority classes more costly [32].
5. **Ensemble Methods:** Using ensemble learning techniques such as Random Forests or Boosting with adjustments for imbalance [33].
6. **Anomaly Detection Techniques:** In certain cases, treating the minority class as anomalies can be effective [34].

We have explored strategies for managing unbalanced datasets, yet we have not addressed how to determine if such imbalances impact model performance. Gaining a thorough understanding of the essential metrics is crucial for this purpose.

8.6 Important Metrics

We have witnessed the effects of an unbalanced dataset. Although visualising decision boundaries can seem helpful for identifying such issues, often the complexity of datasets, with more than two features, renders this visualisation unfeasible. Nevertheless, there is a direct approach to uncovering issues linked to unbalanced datasets: the application of appropriate metrics. Recognising which metric effectively signals an underlying problem is crucial. We will discuss several key metrics. It is important to note that, from a strict mathematical perspective, not all of these are technically 'metrics', but they are significant for our analysis.

- **Confusion Matrix** (described in Section 8.6.2).
- **Sensitivity** and **Specificity** (described in Section 8.6.3).
- **Precision** (described in Section 8.6.4).
- F**-score** (described in Section 8.6.5).
- **Balanced accuracy** (described in Section 8.6.6).
- **Receiving operating characteristic** (ROC) curve (described in Section 8.6.7).

Once you grasp these concepts, you can easily evaluate the performance of your algorithms, especially in terms of their handling of imbalances. An overview of the metrics can be seen schematically in Figure 8.7.

8.6.1 ★ The Notion of Metric

In machine learning, we speak of metrics all the time. Accuracy, mean squared error, mean absolute error, and many more. It is essential to base the idea of a metric on mathematics. Intuitively speaking, a metric is a function that measures the distance between two objects (which could be images, arrays of values, or sound recordings, for example). Take the mean squared error (MSE), given by the following formula:

$$\text{MSE} = \frac{1}{N}\sum_{i=1}^{N}(y_i - z_i)^2 \tag{8.3}$$

where $\mathbf{y} = (y_1, ..., y_N)$ and $\mathbf{z} = (z_1, ..., z_N)$ indicate two vectors in $\mathbb{R}^N$. Equation 8.3 can be interpreted as a *distance* between **y** and **z**. Even if perhaps less intuitive, the error of a classifier (evaluated as $1 - a$, with a the accuracy) can also be interpreted as some form of distance. In fact, imagine that you train a classification model that gives you as output classes 0 or 1 in a binary classification problem with N inputs. Accuracy a can be written as

$$a = \frac{1}{N}\sum_{i=1}^{N}\mathbb{I}(y_i, t_i) \tag{8.4}$$

where y_i and t_i are the predicted and the true class of the ith input respectively, and the function $\mathbb{I}$ is defined as

$$\mathbb{I}(x, z) = \begin{cases} 1 \text{ if } x = z \\ 0 \text{ if } x \neq z \end{cases} \tag{8.5}$$

and thus it simply measures the similarity between the predictions and the true labels. You may wonder why I said that the **error**, in other words, $1 - a$, is a metric and not the accuracy itself. In general, a metric is a function that assigns a real number to each pair of elements of a set in a way that satisfies certain properties. To understand this, we must first define what a metric is.

The notion of a metric appears when discussing and defining **metric spaces** (if you are interested, you should read the exceptional clear and wonderful book by Kolmogorov and Fomin *Elements of the Theory of Functions and Functional Analysis* [35]).

Definition 8.2 A metric on a certain set of objects M (for example, the space that contains all possible arrays of predictions of a model) is a function $d : M \times M \to \mathbb{R}$ that satisfies the following properties.

1. $d(x, x) = 0 \; \forall x \in M$
2. If $x \neq y$ then $d(x, y) > 0 \; \forall x, y \in M$
3. $d(x, z) \leq d(x, y) + d(y, z) \; \forall x, y, z \in M$. This property is called *the triangle inequality*. This inherent trait, whether in physical or abstract notions of distance, suggests that travelling from x to z via y does not offer a shortcut compared to the direct and shortest route.

It is an interesting question if what we call *metrics* in machine learning are really metrics in a more mathematical sense, in other words if they satisfy the three properties above. Let us consider two examples: the Mean Squared Error (MSE) and the accuracy (which is more challenging to prove). For those who are more mathematically inclined or interested in the details, the following two subsections contain more mathematical content, but skipping them will not impact the general understanding of the topic.

8.6.1.1 ★ The MSE is a Metric

To prove that MSE is a metric, we must prove that the MSE function between two vectors $x \in \mathbb{R}^N$ and $y \in \mathbb{R}^N$

$$\text{MSE}(x, y) = \frac{1}{N}\sum_{i=1}^{N}(x_i - y_i)^2 \tag{8.6}$$

satisfies the three properties listed in the previous section.

Lemma 8.1 *The mean squared error (MSE) between two vectors $x \in \mathbb{R}^N$ and $y \in \mathbb{R}^N$ is a metric.*

Proof The proof is divided into three parts, one for each property.

1. $\text{MSE}(x, x) = 0$: this can be easily seen by the definition 8.6.
2. $\text{MSE}(x, y) > 0$ for $x \neq y$: since the MSE is a sum of squares, it is obviously positive.
3. $\text{MSE}(x, y) + \text{MSE}(y, z) \geq \text{MSE}(x, y)$: this can be proven with the Cauchy-Schwarz inequality. In perfect textbook style, I will leave the details to the reader[5].

This concludes the proof. □

8.6.1.2 ★ 1 − *a* is a Metric

Let us start by defining the accuracy in terms of the scalar multiplication of vectors. To keep this section concise, we will focus on binary classification. Extending the discussion to multiple classes would make this section overly lengthy and would go beyond the main scope of this chapter. We will generalise the definition of accuracy of an array of predictions to the concept of *accuracy* of one vector with respect to a second one.

Definition 8.3 (Accuracy in terms of scalar multiplication of vectors) The accuracy $a(x, y)$ of a vector $x \in \mathbb{B}^N$ with respect to a vector $y \in \mathbb{B}^N$ (where $\mathbb{B}$ is the boolean domain, or in other words $\mathbb{B} = \{0, 1\}$) can be written as

$$a(x, y) = \frac{1}{N}(x \cdot y + (\mathbb{1} - x) \cdot (\mathbb{1} - y)) \tag{8.7}$$

where $\mathbb{1} = (1, ..., 1)$ is the unit vector in $\mathbb{B}^N$.

Now feel free to pause and think about the above-mentioned definition. The term $x \cdot y$ is simply the number of elements in the vectors x and y that are both 1. The term $(\mathbb{1} - x) \cdot (\mathbb{1} - y)$ is the number of elements in the two vectors that are both zero. Let us now give another definition first.

[5] Don't you hate when a book says so?

Definition 8.4 (Normalised error function) The normalised error function $\xi(x, y)$ between two vectors $x \in \mathbb{B}^N$ and $y \in \mathbb{B}^N$ is defined as

$$\xi(x, y) = N(1 - a(x, y)) \tag{8.8}$$

where a is the accuracy. To facilitate our discussion, let us prove a lemma before continuing.

Lemma 8.2 *The generalised error function $\xi(x, y)$ between two vectors $x \in \mathbb{B}^N$ and $y \in \mathbb{B}^N$ can be written as*

$$\xi(x, y) = (x - y) \cdot (x - y) = \|x - y\|_2 \tag{8.9}$$

where $\|x\|_2$ is the squared norm, or in other words,

$$\|x\|_2 = \sum_{i=1}^{N} x_i^2 \tag{8.10}$$

Proof

$$\begin{aligned} \xi(x, y) &= N - x \cdot y - (\mathbb{1} - x) \cdot (\mathbb{1} - y) = \\ &= N - x \cdot y - \mathbb{1} \cdot \mathbb{1} + \mathbb{1} \cdot y + \mathbb{1} \cdot x - x \cdot y = \\ &= 2x \cdot y + \mathbb{1} \cdot y + \mathbb{1} \cdot x = \\ &= \{\text{Note that } \mathbb{1} \cdot y = y \cdot y \text{ since } y \in \mathbb{B}^N\} = \\ &= (x - y) \cdot (x - y) = \\ &= \|x - y\|_2^2 \end{aligned} \tag{8.11}$$

That concludes the proof. □

It is easy to prove the following lemma.

Lemma 8.3 *The normalised error $\xi(x, y)$ is a metric.*

Proof To prove the lemma, we need to prove that $\xi(x, y)$ satisfies the three properties of a metric.

1. $\xi(x, x) = 0$: This follows automatically from the lemma 8.2.
2. $\xi(x, y) > 0$ for $x \neq y$: This also comes from the lemma 8.2.
3. $\xi(x, y) + \xi(y, z) > \xi(x, z)$: This can be shown using the Cauchy-Schwarz inequality applied to the 2-norm in the following way

$$\xi(x, y) + \xi(y, z) = \underbrace{\|x - y\|_2^2 + \|y - y\|_2^2 \geq \|y - z\|_2^2}_{\text{Cauchy-Schwarz inequality}} = \xi(x, z) \tag{8.12}$$

This concludes the proof that the normalised error is a metric. □

8.6.2 Confusion Matrix

A confusion matrix, also known as an error matrix, is a two-dimensional table used to evaluate the performance of a classification algorithm. The matrix has two dimensions, one representing the predicted classes (typically the columns) and the other representing the correct classes (typically the rows). In the case of binary classification, the matrix is a 2×2 table, denoted by C. Let us give a proper definition of C.

Definition 8.5 A confusion matrix (C) in a binary classification problem for a model $\mathcal{M}$ is a matrix 2×2, whose entries are defined by

$$C_{ij} = \text{Number of inputs in class } i \text{ that are labelled as class } j \text{ by } \mathcal{M}. \tag{8.13}$$

Let us give a more intuitive definition to make the meaning of C_{ij} clearer.

- C_{11}: the number of inputs that are in class 1 that have been classified as class 1.
- C_{12}: the number of inputs that are classified by the model in class 2 but are really in class 1.
- C_{21}: the number of inputs that the model classified in class 1 but that are actually in class 2.
- C_{22}: the number of inputs that are in class 2 but are really in class 2.

In Figure 8.6 you can see where the C_{ij} are in the matrix (left diagram). Now how

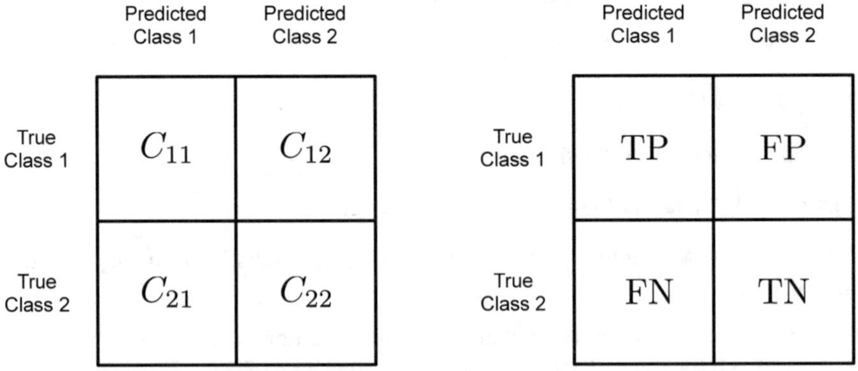

	Predicted Class 1	Predicted Class 2
True Class 1	C_{11}	C_{12}
True Class 2	C_{21}	C_{22}

	Predicted Class 1	Predicted Class 2
True Class 1	TP	FP
True Class 2	FN	TN

Fig. 8.6 On the left a confusion matrix for a binary classification problem is depicted with the C_{ij} elements. They indicate the number of inputs classified in class j (the column) that are actually in class i (the row). On the right the position of the true positives (TP), true negatives (TN), false positives (FP) and false negatives (FN) in the confusion matrix is highlighted.

would C looks in the case depicted in Figure 8.2 for example? All inputs are classified into class 0 although 10 of them are in class 1. So C would be the following

$$C = \begin{pmatrix} 0 & 10 \\ 0 & 1000 \end{pmatrix} \tag{8.14}$$

C in Equation (8.14) should be interpreted as follows:

- The fact that $C_{11} = 0$ tells us that the model is not able to classify any of the inputs in class 1 correctly, and that is bad.
- $C_{22} = 1000$ tells us that the model can correctly classify **all** inputs in class 2.
- The fact that $C_{11} = C_{21} = 0$ tells us that the model classifies **all** inputs in class 2. When this happens, this is a typical symptom of an unbalanced dataset.

A perfect classifier would look like this

$$C = \begin{pmatrix} 10 & 0 \\ 0 & 1000 \end{pmatrix} \tag{8.15}$$

Such a model correctly classifies all inputs in their respective classes. In the case of a binary classification, C_{ij} have special meanings and names (see the right panel of Figure 8.6).

- C_{11} is called **True Positives** (TP), the number of inputs that are in class 1 that are actually in class 1.
- C_{22} is called **True Negatives** (TN),the number of inputs that are in class 2 that are actually in class 2.
- C_{21} is called **False Negatives** (FN) (also called **type II error**), the number of inputs in class 2 that are incorrectly classified as class 1.
- C_{12} is called **False Positives** (FP) (also called **type I error**), the number of inputs in class 1 that are incorrectly classified as class 2.

The names "positives" and "negatives" come from cases where the classes indicate the presence (or absence) of a certain condition (as in medicine). For example, TP could indicate the number of sick patients correctly identified and FP the number of healthy patients incorrectly classified as sick. But in general, what is positive and what is negative is arbitrary and depends on the classes. To gain an intuitive understanding of these quantities, consider the following example. Imagine a study that tries to identify people with a disease. Each person is sick or not. Each person can be classified as positive (the person has the disease) or negative (the person does not have the disease). In this case, we have the following.

- TP: Sick people correctly identified as sick.
- FP: Healthy people incorrectly identified as sick.
- TN: Healthy people correctly identified as healthy.
- FN: Sick people incorrectly identified as healthy.

By evaluating the confusion matrix, it is possible to easily check if a model is classifying all inputs (or the majority) in one single class (for example, classifying everyone as sick or healthy). If that happens, this is a clear symptom of an unbalanced dataset. It is always a good idea to check the confusion matrix in a classification problem, as it provides quite some important information on the performance of a classifier for each class. In the case of k classes, the matrix will be a $k \times k$ matrix, but the elements will have the same meaning. That is, C_{ij} will indicate the number of inputs classified in class j that are actually in class i, where i and j can assume all values between 1 and k (the k classes).

The name **confusion** matrix, comes from the idea that this matrix gives an assessment of how much the model is confused about its prediction. A diagonal matrix (in other words, with $C_{ij} = 0$ for $i \neq j$) indicates a model that is *not* confused at all, while one with the diagonal elements equal to zero is a completely confused model (one that cannot predict better than one that throws coins in the air for its prediction).

8.6.3 Sensitivity and Specificity

Sensitivity (also called **recall, hit rate** or **true positive rate**) and **specificity** measure the ability of a model to detect the presence of absence of a condition. They can only be defined in the case of binary classification. Sensitivity is defined by

$$\text{Sensitivity} = \frac{\text{TP}}{\text{P}} = \frac{\text{TP}}{\text{TP} + \text{FN}} \tag{8.16}$$

Specificity (also called **selectivity** or **true negative rate**) is defined by

$$\text{Specificity} = \frac{\text{TN}}{\text{N}} = \frac{\text{TN}}{\text{TN} + \text{FP}} \tag{8.17}$$

Info Sensitivity and Specificity with Unbalanced Datasets

When you are dealing with an unbalanced datasets, if the model ignores the minority class, this will be reflected in one of the two metrics, sensitivity or specificity, to be very low. Another way of checking if you are having issues due to the unbalance of the data, is to check those two quantities. In a good classifier, both should be high.

Sensitivity measures the proportion of elements in class 1 that the model can accurately identify. If class 1 indicates patients with a disease, sensitivity will indicate how many sick patients can be correctly identified within all sick patients. Or said in an even different way, how many of all sick patients can the model identify correctly. A model that has a very high sensitivity, can be used successfully to rule-out a disease for example. Sensitivity is also called **recall, hit rate** or **true positive rate**. Specificity, on the other hand, indicates what percentage of all elements in class 2 are correctly identified. Specificity is also called **selectivity** or **true negative rate**. Note that the two can be exchanged, since what is positive and what is negative is a decision taken in the problem definition. So, basically, they assess the same information for both classes.

Info Origin of the Names Specificity and Sensitivity

It is interesting to ask why these two quantities are called specificity and sensitivity. The origin comes from their use in medicine. In fact, specificity, from an etymological point of view, means *condition of being specific* [36], where *specific* means *having a specific quality* from Latin *specificus*, or constituting a kind or sort. This term was used in a medical context to indicate the presence or absence of a disease or condition (having a specific quality, a disease or the absence of it). They were originally immunological concepts, closely associated with the diagnosis of syphilis [37] and used as far back as the 1900s. Again, it is important to note that those two concepts are interchangeable. The switch of labels results in the sensitivity and specificity to switch roles.

8.6.4 Precision

There is more important information when dealing with binary classification. How many of the inputs identified as *positives* (in class 1 for example) within all positively identified inputs are correct (truly positive)? **Precision** answers exactly this question.

$$\text{Precision} = \frac{\text{TP}}{\text{TP} + \text{FP}} \tag{8.18}$$

Precision is often used together with specificity and sensitivity. Note that precision can only be evaluated if $\text{TP} + \text{FP} \neq 0$. Since both can only be positive, that means that TP and FP must be both different than zero. $\text{TP} = 0$ and $\text{FP} = 0$ can only happen if the model classifies all inputs in class 2 (see Figure 8.6). In this case, precision cannot be evaluated. When $\text{TP} + \text{FP} = 0$, the python function `sklearn.metrics.precision_score` will return 0 but also raises a warning since the metric is not defined [38].

8.6.5 F_β-score

The metrics discussed so far are always specific to one single class. They assess how good (or bad) a model is when predicting one single class. In general, one wants to find a balance between the two classes, or in other words, one wants a model that can perform reasonably well on both classes. The following metrics combine multiple class-specific metrics to assess how balanced they are.

The F_β-score is a metric that is often used to assess the performance of a classifier with β a real number. In all practical applications, there are three values of β that are used: 0.5, 1 and 2. The general formula for β is

$$F_\beta = (1+\beta^2)\frac{\text{precision} + \text{recall}}{\beta^2\text{precision} + \text{recall}} \tag{8.19}$$

For $\beta = 0.5$ recall is weighted less than precision. For $\beta = 2$ the opposite. For $\beta = 1$ you get the harmonic average between recall and precision. With some algebra you can see that

$$F_1 = \frac{2}{\text{recall}^{-1} + \text{precision}^{-1}} \tag{8.20}$$

Which value of β you should choose depends strongly on your problem and what you are interested in. F_1 is a metric that weighs the same recall and precision. In general, the highest possible value of the F_β metric is 1.0, indicating perfect precision and recall (equal to 1). If recall and precision are exactly zero at the same time, F_β is not defined, since division by zero is not possible. This metric only makes sense for recall and precision different from zero.

Warning F_β score and division by zero

To evalute F_β one has to calculate precision and recall. Recall that

$$\text{precision} = \frac{\text{TP}}{\text{TP} + \text{FP}} \tag{8.21}$$

and

$$\text{recall} = \frac{\text{TP}}{\text{TP} + \text{FN}} \tag{8.22}$$

What happens if TP+FP or TP+FN are equal to zero in practical applications? If that happens, F_β cannot be evaluated. If you are using the python scikit-learn function `sklearn.metrics.fbeta_score` this is checked for you. In fact, if TP+FP = 0 or TP+FN = 0 then the function will return a value for F_β equal to zero and at the same time it will raise a `UndefinedMetricWarning` [39]. So do not take numerical values at face value, but check if they make sense (in Python you may check warnings as the above mentioned one).

It is important to highlight another aspect of the F_β score. Both precision and recall are metrics that give information about the *positive* class (check the formulas again). This metric will not tell you much about how the model behaves with the other class (the *negative* one). This comes from the fact that historically the *positive* class was of greatest interest (for example, sick patients).

8.6.6 Balanced Accuracy

Balanced accuracy is nothing more than the average of sensitivity and specificity and is often indicated with a_B.

$$a_B = \frac{\text{sensitivity} + \text{specificity}}{2} \tag{8.23}$$

Balanced accuracy is a metric used in classification, especially when dealing with imbalanced datasets. It is the arithmetic mean of sensitivity (true positive rate) and specificity (true negative rate), which offers a more insightful performance measure when the class distribution is skewed: a balanced accuracy of 1 indicates perfect classification, while a value of 0.5 suggests no better performance than random guessing.

8.6.7 Receiving Operating Characteristic (ROC) Curve

Another widely used way to assess the performance of classifiers is the receiving operating characteristic (ROC) curve. Suppose that we have a model that classifies some input into two classes (the ROC curve can only be defined for binary classification). A model can give two kinds of output.

1. Discrete output: the model will give the class as output.
2. Probabilistic output: the model gives the probability that the input is in one of the two classes.

A ROC curve can only be created for models that fall in the second category, as it will become clear very soon. In this section, we then assume that the model will predict the probability $p(1|x_i)$, that is, the probability of having class 1 given the input x_i. Normally, you assign the input x_i to class 1 (recall that we are talking about a binary classification, so we have only two classes that we will indicate with 0 and 1) if $p(1|x_i) > \theta$, with $\theta \in \mathbb{R}$ some threshold, typically chosen as 0.5. The choice of θ is somewhat arbitrary. It could well be chosen as 0.3 or 0.7 for example. Now, as you can easily imagine, the values for the TP, TN, FP and FN will depend on θ. The receiving operating characteristic (ROC) curve is obtained by drawing TPR, that is, the probability of predicting 1 if the real class is 1 versus FPR, that is the probability of predicting 1 if the real class is 0 varying θ from 0 to 1. Let us give an example. Consider the *iris* dataset [40, 41].

Info The Iris Dataset

The data set consists of 50 samples from three species of Iris: Iris setosa, Iris virginica and Iris versicolor. For each species four characteristics were measured: length and width of the sepals and petals. In Figure 8.8 you can see what petals and sepals are.

Fig. 8.8 Petals and sepals in a flower (Photo Eric Guinther CC BY-SA 3.0).

The dataset comprises three categories, each with 50 samples, representing different types of iris plants. One category can be linearly distinguished from the other two, but the other two categories are not linearly distinguishable from one another. A nice article on the dataset that contains a lot of information is that by Unwin and Leinmann [42]. I suggest you read it, it is quite interesting.

Now if we take this dataset and train a support vector classifier (don't worry if you do not know the model, this will not impact your understanding) on only two classes of the three present in the dataset, we can plot the ROC curve as described above. The model will output a probability $\hat{y}_i$ for each input i. We obtain the ROC curve by classifying the inputs into two classes by varying the threshold θ. The result can be seen in Figure 8.9. Some important points about the ROC.

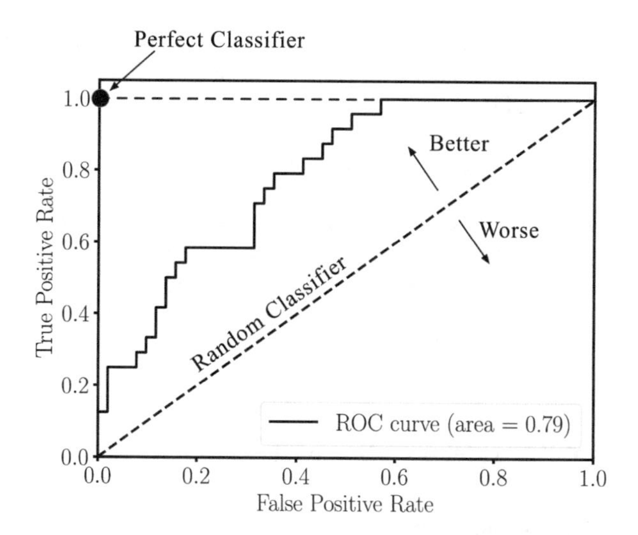

Fig. 8.9 The ROC curve obtained by varying the threshold θ with a support vector classifier trained on the iris dataset. The dashed diagonal line is the line you would obtain with a completely random classifier (one that would give you 50% of accuracy). The more the ROC curve is above the dashed diagonal line, the better the classifier is.

- The diagonal dashed line is the one you would obtain with a random classifier, one that would give you 50% accuracy. Think about it, on the diagonal line TPR = FPR
$$\text{TPR} = \frac{\text{TP}}{\text{P}} = \frac{\text{FP}}{\text{N}} = \text{FPR} \tag{8.24}$$
that means that one **always** identifies the same number of positives (between all positives), TP / P, as the same amount of **false** positives. This can only happen if the ratio is 50/50, or, in other words, if one makes as many mistakes as many correct classifications.
- The more the ROC curve is above the dashed line, the better the classifier is.
- the point (0,1) indicates a perfect classifier. In fact, FPR = 0, means that TP = P. In other words, the model identifies all positive inputs correctly. And FPR = 0 means FP = 0, or, in other words, the model does not make mistakes.
- Any classifier that appears in the lower right triangle performs worse than random guessing. But it is enough to switch all predicted classes (all predicted 0 becomes 1 and all 1 becomes 0) so that its TP becomes FN, and its FP becomes TN. Thus, a classifier appearing in the lower right triangle can be transformed into one that appears in the upper left triangle [43]. By switching the predictions, every point in the lower right triangle will be reflected to one in the upper left triangle symmetrical with respect to the point $(0.5, 0.5)$.

To extract a single number from the ROC curve, the area under the ROC curve, indicated with AUC (**A**rea **U**nder the **C**urve), is often calculated, and given as a measurement of how close the ROC curve is to a perfect classifier. AUC = 1 indicates a perfect classifier, AUC = 0.5 a completely random one.

Note that if you train a classifier and you get some predictions, this will correspond to one single point in the so-called *ROC space*, or in other words in the space given by $\{0, 1\} \times \{0, 1\}$ for TPR and FPR.

Info Probability interpretation of the AUC

The AUC is the probability that given one randomly selected input in class 1 (positives) and one randomly selected negative input in class 0 (negatives), the model (the classifier) will be able to tell which is which. The mathematical explanation can be found in Section 8.6.8.

8.6.8 ★ Probability Interpretation of the AUC

To understand the probability interpretation of the AUC it is necessary to introduce some additional mathematical notation. Suppose that we have a classifier that outputs the probability (which we will indicate with X) of an input that belongs to a certain class. We consider X as a random variable. Given a threshold $T \in \mathbb{R}$, the input is classified as in class 1 (positive) if $X > T$, and in class 0 (negative) if $X < T$. Indicating with $f_1(x)$ the probability density if the instance *actually* belongs to class 1, and with $f_0(x)$ if the instance belongs to class 0 we can write the TPR(T) as

$$\text{TPR(T)} = \int_T^\infty f_1(x)dx \tag{8.25}$$

and FPR(T) as

$$\text{FPR(T)} = \int_{-\infty}^T f_0(x)dx \tag{8.26}$$

The ROC curve is a parametric plot of TPR(T) vs FPR(T) for T varying from $-\infty$ to ∞. How the curves are really parameterised is completely irrelevant for the discussion, and it depends strongly on the model, the data, etc. Now we can write the AUC

$$\begin{aligned}
\text{AUC} = \int_{-\infty}^{\infty} \text{TPR}(\text{FPR})d(\text{FPR}) = \\
= \{\text{FPR} \rightarrow \text{FPR}(T) \Rightarrow \\
\text{TPR}(\text{FPR})d(\text{FPR}) = \text{TPR}(\text{FPR}(T))\text{FPR}'(T)d(T)\} = \\
= \int_{-\infty}^{\infty} \text{TPR}(\text{FPR}(T))\text{FPR}'(T)dT
\end{aligned} \tag{8.27}$$

Now note that from Equation 8.26

$$\text{FPR}'(\text{T}) = f_0(T) \tag{8.28}$$

Therefore

$$
\begin{aligned}
\text{AUC} &= \int_{-\infty}^{\infty} f_1(T)dT \int_{-\infty}^{T} f_0(T')dT' = \\
&= \{\text{Note thta in this integral } T \geq T'\} = \\
&= \int_{-\infty}^{\infty}\int_{-\infty}^{\infty} I(T \geq T') f_1(T) f_0(T')dTdT' = \\
&= P(X_1 > X_0)
\end{aligned} \tag{8.29}
$$

Where $I(x)$ is one if $x \geq 0$ and 0 otherwise. The interpretation of this result is that the AUC is the probability that given one randomly selected input in class 1 (positives) and one randomly selected negative input in class 0 (negatives), the model (the classifier) will be able to tell which is which.

Info ■ ROC Curve Beyond Binary Classification

Extending the ROC curves for multi-class problems, is cumbersome and difficult. There are two common approaches.

1. Average over all pairs of AUC values obtained by comparing all classes two at a time [44].
2. Volume under the surface [45, 46].

Those approaches are rarely used, and the interested reader is directed to the cited papers for more information.

References

1. Gilles Cohen, Mélanie Hilario, Hugo Sax, Stéphane Hugonnet, and Antoine Geissbuhler. Learning from imbalanced data in surveillance of nosocomial infection. *Artificial intelligence in medicine*, 37(1):7–18, 2006.
2. Haibo He and E.A. Garcia. Learning from Imbalanced Data. *IEEE Transactions on Knowledge and Data Engineering*, 21(9):1263–1284, September 2009.
3. Yanmin Sun, Andrew K. C. Wong, and Mohamed S. Kamel. CLASSIFICATION OF IMBALANCED DATA: A REVIEW. *International Journal of Pattern Recognition and Artificial Intelligence*, 23(04):687–719, June 2009.
4. Salvador García, Francisco Herrera, and Julián Luengo. *Data Preprocessing in Data Mining*. Number 72 in Intelligent Systems Reference Library. Springer International Publishing : Imprint: Springer, Cham, 1st ed. 2015 edition, 2015.
5. Paula Branco, Luis Torgo, and Rita Ribeiro. A Survey of Predictive Modelling under Imbalanced Distributions, May 2015. arXiv:1505.01658 [cs].
6. Shweta Sharma, Anjana Gosain, and Shreya Jain. A Review of the Oversampling Techniques in Class Imbalance Problem. In Ashish Khanna, Deepak Gupta, Siddhartha Bhattacharyya, Aboul Ella Hassanien, Sameer Anand, and Ajay Jaiswal, editors, *International Conference on Innovative Computing and Communications*, volume 1387, pages 459–472. Springer Singapore, Singapore, 2022. Series Title: Advances in Intelligent Systems and Computing.

7. Georgios Douzas and Fernando Bacao. Self-Organizing Map Oversampling (SOMO) for imbalanced data set learning. *Expert Systems with Applications*, 82:40–52, October 2017.
8. Tian-Yu Liu. EasyEnsemble and Feature Selection for Imbalance Data Sets. In *2009 International Joint Conference on Bioinformatics, Systems Biology and Intelligent Computing*, pages 517–520, Shanghai, China, 2009. IEEE.
9. V. Bolón-Canedo, N. Sánchez-Maroño, A. Alonso-Betanzos, J.M. Benítez, and F. Herrera. A review of microarray datasets and applied feature selection methods. *Information Sciences*, 282:111–135, October 2014.
10. Classification on imbalanced data - tensorflow core. `https://www.tensorflow.org/tutorials/structured_data/imbalanced_data`. (Accessed on 01/08/2023).
11. Roweida Mohammed, Jumanah Rawashdeh, and Malak Abdullah. Machine learning with oversampling and undersampling techniques: overview study and experimental results. In *2020 11th international conference on information and communication systems (ICICS)*, pages 243–248. IEEE, 2020.
12. Mayuri S Shelke, Prashant R Deshmukh, and Vijaya K Shandilya. A review on imbalanced data handling using undersampling and oversampling technique. *Int. J. Recent Trends Eng. Res*, 3(4):444–449, 2017.
13. B Santoso, H Wijayanto, KA Notodiputro, and B Sartono. Synthetic over sampling methods for handling class imbalanced problems: A review. In *IOP conference series: earth and environmental science*, volume 58, page 012031. IOP Publishing, 2017.
14. Ahmad S Tarawneh, Ahmad B Hassanat, Ghada A Altarawneh, and Abdullah Almuhaimeed. Stop oversampling for class imbalance learning: A review. *IEEE Access*, 2022.
15. Rushi Longadge and Snehalata Dongre. Class imbalance problem in data mining review. *arXiv preprint arXiv:1305.1707*, 2013.
16. Julio Hernandez, Jesús Ariel Carrasco-Ochoa, and José Francisco Martínez-Trinidad. An empirical study of oversampling and undersampling for instance selection methods on imbalance datasets. In *Iberoamerican Congress on Pattern Recognition*, pages 262–269. Springer, 2013.
17. Roweida Mohammed, Jumanah Rawashdeh, and Malak Abdullah. Machine learning with oversampling and undersampling techniques: Overview study and experimental results. In *2020 11th International Conference on Information and Communication Systems (ICICS)*, pages 243–248, 2020.
18. Nathalie Japkowicz and Shaju Stephen. The class imbalance problem: A systematic study. *Intelligent data analysis*, 6(5):429–449, 2002.
19. Le Wang, Meng Han, Xiaojuan Li, Ni Zhang, and Haodong Cheng. Review of classification methods on unbalanced data sets. *IEEE Access*, 9:64606–64628, 2021.
20. Ray Marie Tischio and Gary M Weiss. Identifying classification algorithms most suitable for imbalanced data. *Dept. Comput. Inf. Sci., Fordham Univ., The Bronx, NY, USA, Tech. Rep*, 2019.
21. Justin M Johnson and Taghi M Khoshgoftaar. Survey on deep learning with class imbalance. *Journal of Big Data*, 6(1):1–54, 2019.
22. The imbalanced-learn developers. Over-sampling - Version 0.10.1. `https://imbalanced-learn.org/stable/over_sampling.html`. (Accessed on 01/01/2023).
23. N. V. Chawla, K. W. Bowyer, L. O. Hall, and W. P. Kegelmeyer. SMOTE: Synthetic Minority Over-sampling Technique. *Journal of Artificial Intelligence Research*, 16:321–357, June 2002.
24. The imbalanced-learn developers. SMOTE — Version 0.10.1. `https://imbalanced-learn.org/stable/references/generated/imblearn.over_sampling.SMOTE.html`. (Accessed on 01/13/2023).
25. The imbalanced-learn developers. Under-sampling - Version 0.10.1. `https://imbalanced-learn.org/stable/under_sampling.html`. (Accessed on 01/01/2023).
26. Nitesh V. Chawla, Kevin W. Bowyer, Lawrence O. Hall, and W. Philip Kegelmeyer. Smote: Synthetic minority over-sampling technique. *Journal of Artificial Intelligence Research*, 16:321–357, 2002.
27. Nathalie Japkowicz and Shaju Stephen. The class imbalance problem: A systematic study. *Intelligent Data Analysis*, 6:429–449, 2002.

28. Ting Jo, Ng, and Kotagiri Ramamohanarao. Class imbalances versus small disjuncts. *SIGKDD Explor. Newsl.*, 6:40–49, 2004.
29. Nitesh V. Chawla, Kevin W. Bowyer, Lawrence O. Hall, and W. Philip Kegelmeyer. Smote: Synthetic minority over-sampling technique. In *Journal of Artificial Intelligence Research*, 2002.
30. Hui Han, Wen-Yuan Wang, and Bing-Huan Mao. Borderline-smote: A new over-sampling method in imbalanced data sets learning. *International Conference on Intelligent Computing*, pages 878–887, 2005.
31. Haibo He, Yang Bai, Edwardo A. Garcia, and Shutao Li. Adasyn: Adaptive synthetic sampling approach for imbalanced learning. In *IEEE International Joint Conference on Neural Networks (IEEE World Congress on Computational Intelligence)*, pages 1322–1328, 2008.
32. Charles Elkan. The foundations of cost-sensitive learning. In *International Joint Conference on Artificial Intelligence*, pages 973–978, 2001.
33. Chao Chen, Andy Liaw, and Leo Breiman. Using random forest to learn imbalanced data. *University of California, Berkeley*, 2004.
34. Varun Chandola, Arindam Banerjee, and Vipin Kumar. Anomaly detection: A survey. *ACM Computing Surveys (CSUR)*, 41(3):1–58, 2009.
35. A. N. Kolmogorov and S. V. Fomin. *Elements of the Theory of Functions and Functional Analysis*. Dover Publications, 1961.
36. specificity | Etymology, origin and meaning of specificity by etymonline. `https://www.etymonline.com/word/specificity`. (Accessed on 26/11/2023).
37. Nicholas Binney, Christopher Hyde, and Patrick M. Bossuyt. On the Origin of Sensitivity and Specificity. *Annals of Internal Medicine*, 174(3):401–407, March 2021. Publisher: American College of Physicians.
38. sklearn.metrics.precision_score. `https://scikit-learn/stable/modules/generated/sklearn.metrics.precision_score.html`. (Accessed on 26/11/2023).
39. sklearn.metrics.fbeta_score. `https://scikit-learn/stable/modules/generated/sklearn.metrics.fbeta_score.html`. (Accessed on 26/11/2023).
40. Ronald A Fisher. The use of multiple measurements in taxonomic problems. *Annals of eugenics*, 7(2):179–188, 1936.
41. Edgar Anderson. The species problem in iris. *Annals of the Missouri Botanical Garden*, 23(3):457–509, 1936.
42. Antony Unwin and Kim Kleinman. The Iris Data Set: In Search of the Source of Virginica. *Significance*, 18(6):26–29, December 2021.
43. Tom Fawcett. An introduction to ROC analysis. *Pattern Recognition Letters*, 27(8):861–874, June 2006.
44. David J Hand and Robert J Till. A simple generalisation of the area under the roc curve for multiple class classification problems. *Machine learning*, 45:171–186, 2001.
45. Douglas Mossman. Three-way rocs. *Medical Decision Making*, 19(1):78–89, 1999.
46. César Ferri, José Hernández-Orallo, and Miguel Angel Salido. Volume under the roc surface for multi-class problems. In *European conference on machine learning*, pages 108–120. Springer, 2003.

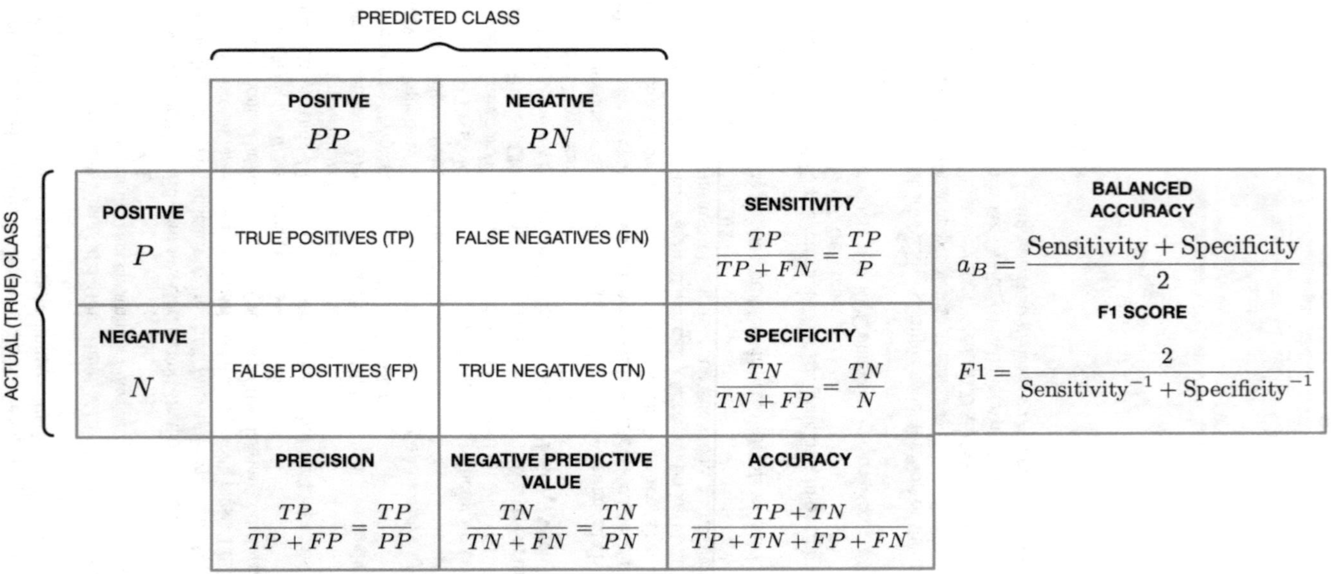

		PREDICTED CLASS			
		POSITIVE PP	**NEGATIVE** PN		
ACTUAL (TRUE) CLASS	**POSITIVE** P	TRUE POSITIVES (TP)	FALSE NEGATIVES (FN)	**SENSITIVITY** $\frac{TP}{TP+FN} = \frac{TP}{P}$	**BALANCED ACCURACY** $a_B = \frac{\text{Sensitivity} + \text{Specificity}}{2}$
	NEGATIVE N	FALSE POSITIVES (FP)	TRUE NEGATIVES (TN)	**SPECIFICITY** $\frac{TN}{TN+FP} = \frac{TN}{N}$	**F1 SCORE** $F1 = \frac{2}{\text{Sensitivity}^{-1} + \text{Specificity}^{-1}}$
		PRECISION $\frac{TP}{TP+FP} = \frac{TP}{PP}$	**NEGATIVE PREDICTIVE VALUE** $\frac{TN}{TN+FN} = \frac{TN}{PN}$	**ACCURACY** $\frac{TP+TN}{TP+TN+FP+FN}$	

Fig. 8.7 An overview of important metrics as functions of the quantities TP, TN, FP, FN.

Chapter 9
Hyper-parameter Tuning

Each moment spent searching is also a moment spent finding.
Paul Cohello

Hyper-parameters can be loosely defined as those parameters that are not changed during the training process. For example, number of layers in a FFNN, number of neurons in each layer, activation functions, learning rate, and so on. In this chapter I will discuss the problem of finding the best hyper-parameters to get the best results from your models. Doing this is called hyper-parameter tuning. I will first describe what a black-box optimisation problem is and how those classes of problems relate to hyper-parameter tuning. I will discuss the two most known methods to tackle these kind of problems: grid search and random search. You will understand, with examples, which one works under which conditions, and a few tricks that are very helpful, such as *coarse to fine* optimisation and *sampling on a logarithmic scale*. At the end of the chapter, you should know what hyper-parameter tuning is, why it is important, and how it works.

9.1 Introduction

Model learning and, in particular, neural networks have many parameters that can be changed when designing and training them. In the context of neural networks, those could be the number of layers, activation functions, number of neurons, type of neurons, loss function, learning rate, number of epochs, batch size, and many more. But how should we know which values for each parameter are the best? There is no easy answer and different sets of values must be tested, since they strongly depend on the data you have, the size of the dataset and changing one parameter may influence the effectiveness of the others. The process of searching for the best parameters for a given problem and dataset is called *hyper-parameter tuning*. A hyper-parameter can be defined (in a slightly loose way) as follows.

Definition 9.1 A **hyper-parameter** is a parameter that is used in the design of machine learning algorithms or the learning process and that is not changed by the training process and the training data.

U. Michelucci, *Fundamental Mathematical Concepts for Machine Learning in Science*,
https://doi.org/10.1007/978-3-031-56431-4_9

For example, the weights in a neural network are not hyper-parameters, since they change during training. The number of layers in a feed-forward neural network, on the other side, will remain constant during the entire training phase, and thus it is an hyper-parameter.

This chapter deals with the process of finding the best values for the hyper-parameters for a given problem.

9.2 Black-box Optimisation

The problem of hyper-parameter tuning is just a subclass of a much more general type of problem: black-box optimisation. A black-box function $f(x)$

$$f(x) : \mathbb{R}^n \rightarrow \mathbb{R} \tag{9.1}$$

is a function whose analytic form is unknown. $f(x)$ can be evaluated to obtain its value for all the values of x on which it is defined, but no other information, such as its gradient, can be obtained analytically. Generally, we talk about global optimisation of a black-box function (sometimes called a black-box problem) when we try to find the maximum or minimum of $f(x)$ given certain constraints. Here are some examples of this kind of problems:

- Finding the hyper-parameter for a given machine learning model that maximise the chosen metric.
- Finding a maximum or minimum of a function that can only be evaluated numerically or with code that we cannot look at. In some industry contexts, there is legacy code that is very complicated and there are some functions that must be maximised based on its outcome.
- Finding the best place to drill for oil. Your function would be how much oil you can find and x is your location.
- Finding the best combination of parameters for situations that are too complex to model, for example, when launching a rocket in space: how to optimise amount of fuel, diameter of each stage of the rocket, precise trajectory, etc.

This is a very fascinating class of problems that has produced a staggering number of solutions. We will look at the two most important strategies to solve them: grid search and random search.

First, we should ask ourselves why finding the best hyper-parameters for machine learning, especially for neural networks, is a black-box problem. Since we cannot calculate information like the gradients of our network output with respect to the hyper-parameters (for example, the number of layers in an FFNN), especially when using complex optimisers or custom functions, we need other approaches to be able to find the best hyper-parameters that maximise the chosen optimising metric. In this chapter, our black-box function f will be our neural network model (including things like the optimiser, the form of the cost function, etc.) that gives as output a metric given the hyper-parameters as input. In this case, x will denote the hyper-parameters.

The problem may seem easy to solve, so why not try all the possibilities? If you are working on a problem and training your model takes a week, this may present a challenge. Since you will typically have several hyper-parameters, trying all possibilities will not be feasible. Let us give an example and calculate some numbers. Let us suppose that we are training a neural network with several layers. We may decide to consider the following hyper-parameters to see which combination works better (if you do not know what each parameter is, do not worry, you do not need to understand them to follow the discussion):

- Learning rate: let's suppose we want try the values $n10^{-4}$ for $n = 1, \ldots, 10^2$ (100 values)
- Regularization parameter: 0, 0.1, 0.2, 0.3, 0.4 and 0.5 (6 values)
- Choice of optimizer: GD, RMSProp or Adam (3 values)
- Number of hidden layers: 1,2,3,5 and 10 (5 values)
- Number of neuron in the hidden layers: 100, 200 and 300 (3 values)

Consider that you will need to train your network

$$100 \times 6 \times 3 \times 5 \times 3 = 27000 \tag{9.2}$$

times if you want to test all possible combinations. If your training takes 5 minutes, you will need 13.4 weeks of computing time. If training takes hours or days, this will no longer be possible. If training takes one day, for example, you will need 73.9 years of computing time to try all possibilities. Most of the hyper-parameter choices will come from experience; for example, you can always safely use Adam, since it is the better optimiser out there (in almost all cases). But you will not be able to avoid trying to tune other parameters such as the number of hidden layers or the learning rate.

9.2.1 Notes on Black-box Functions

Black-box functions are usually classified in two main classes:

- Cheap functions: functions that can be evaluated thousands of times. For example, evaluating the function could be very fast, or it could be cheap from an economic point of view.
- Costly functions: functions that can only be evaluated a few times, usually less than 10-100 times. A classical example is how much oil we get if we drill at a given location. Getting this information costs time and, especially, a lot of money, so drilling to get the amount of oil in a specific location can be done only a few times before it becomes, from an economic point of view, unsustainable anymore.

If the black-box function is cheap, then the choice of the optimisation method is not critical. For example, when we can evaluate the gradient with respect to the x numerically, or simply search for the maximum by evaluating the functions on a high number of points. If the function is costly, we need much smarter approaches.

Note that neural networks are almost always, especially in the deep learning world, costly functions. For costly functions, we need to find methods that solve our problem with the smallest number of evaluations possible.

9.3 The Problem of Hyper-parameter Tuning

Before looking at how we can find the best hyper-parameters we need to quickly go back to neural networks and discuss what we can tune. Typically, when talking about hyper-parameters, beginners think only of numerical parameters, such as learning rate. Recall that also the following can be varied to see if you can get better results (we will use neural networks, since in this context there are a much higher number of hyper-parameters to use as examples):

- Number of epochs: sometimes simply training your network longer will give you better results.
- Choice of optimiser: you can try to choose a different optimiser. If you are using plain gradient descent, you may try Adam and see if you get better results.
- Varying the regularisation method: there are several ways of applying regularisation. Varying the method may be worthwhile.
- Choice of activation function: although the activation function used often for neurons in hidden layers is ReLU, others may work much better. Trying, for example, sigmoid or Swish[1] [1] may help you get better results.
- Number of layers and number of neurons in each layer: try different configurations. Try layers with different number of neurons for example.
- Learning rate decay methods: try (if you are not using optimisers that do that already) different learning rate decay methods.
- Mini-batch size: vary the size of mini-batches. When you have few data, you can use batch gradient descent; when you have a lot of data, mini-batches are more efficient.
- Weight initialisation methods.

As you can see, there are quite some aspects of your deep learning models that you can change and consider that the list is not exhaustive! Let us now classify the parameters that we can adjust in our models into the following three categories:

1. Parameters that are **continuous real numbers**, or in other words that they can assume any value. Example: learning rate, regularisation parameter.
2. Parameters that are **discrete** but can theoretically assume an infinite number of values. Example: number of hidden layers, number of neurons in each layer, or number of epochs.
3. Parameters that are **discrete** and can only assume a **finite number of possibilities**. Example: optimiser, activation function, learning rate decay method.

[1] Swish is an activation function used in neural networks, defined as $f(x) = x \cdot \sigma(\beta x)$, where $\sigma(x)$ is the sigmoid function, designed to improve model performance by combining the properties of linear and non-linear functions.

For category 3 there is not much to do except try all possibilities. They typically will change completely the model itself and therefore are impossible to model their effect, therefore making a test the only possibility. This is also the category where experience helps the most. It is widely known that Adam optimiser is almost always the best choice, for example, so you may concentrate your effort elsewhere, especially at the beginning. For categories 1 and 2 is a bit more difficult, and we will need to find some smart ideas to find the best values.

9.4 Sample Black-box Problem

To try our hand at solving a black-box problem, let us create a "fake" black-box problem. The problem is the following: find the maximum of the function $f(x)$ in the range $[0, 80]$ given by the formula

$$\begin{cases} g(x) = \cos\frac{x}{4} - \sin\frac{x}{4} - \frac{5}{2}\cos\frac{x}{2} + \frac{1}{2}\sin\frac{x}{2} \\ h(x) = -\cos\frac{x}{3} - \sin\frac{x}{3} - \frac{5}{2}\cos\frac{2}{3}x + \frac{1}{2}\sin\frac{2}{3}x \\ f(x) = 10 + g(x) + \frac{1}{2}h(x) \end{cases} \tag{9.3}$$

pretending not to know the formula itself. The formula will let us check our results, but we will pretend that it is unknown. You may wonder why the choice of such a complicated formula. I wanted to have something with a few maxima and minima to give an idea how the methods are working on a non-trivial example. In Figure 9.1 you can see how $f(x)$ looks. The maximum is at an approximate value $x = 69.18$ and has a value of 15.027. Our challenge is to find this maximum in the most efficient way possible, without knowing anything about $f(x)$ except its value at any point we want. When we say "efficient" we mean, of course, with the smallest number of evaluations possible.

9.4.1 Grid Search

The first method we look at, grid search, is also less "intelligent". Grid search entails simply trying the function at regular intervals and seeing for which x the function $f(x)$ assumes the highest value. In this example, we want to find the maximum of the function $f(x)$ between two x values x_{min} and x_{max}. We simply take n points equally spaced between x_{min} and x_{max} and evaluate the function at these points. We define a vector of points.

$$\mathbf{x} = \left(x_{\text{min}}, x_{\text{min}} + \frac{\Delta x}{n}, \ldots, x_{\text{min}} + (n-1)\frac{\Delta x}{n}\right) \tag{9.4}$$

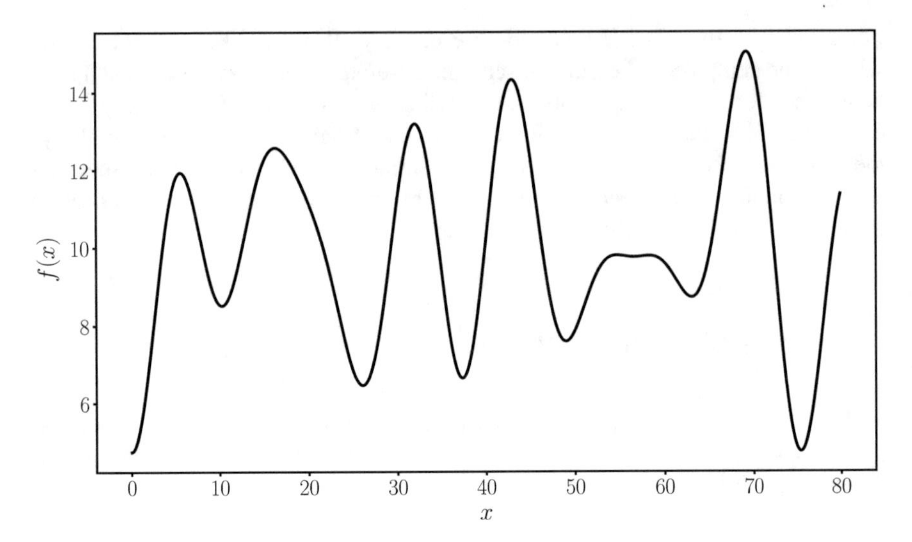

Fig. 9.1 A plot of the function $f(x)$ as described in the text.

where we defined $\Delta x = x_{\max} - x_{\min}$. Then we evaluate the function $f(x)$ at those points, obtaining a vector $\mathbf{f}$ of values

$$\mathbf{f} = \left(f(x_{\min}), f\left(x_{\min} + \frac{\Delta x}{n}\right), \ldots, f\left(x_{\min} + (n-1)\frac{\Delta x}{n}\right)\right) \tag{9.5}$$

the estimate of the maximum $\tilde{f}$ will then be

$$\tilde{f} = \max_{0 \le i \le n-1} f_i \tag{9.6}$$

and supposing the maximum is found at $i = \tilde{i}$ we will also have for its location

$$\tilde{x} = x_{\min} + \frac{\tilde{i}\Delta x}{n} \tag{9.7}$$

Now, as you may imagine the more points you use the more accurate will be your maximum estimation. The problem is that, if the evaluation of $f(x)$ is costly, you will not be able to take as many points as you might like. You will need to find a balance between number of points and accuracy. Let us make an example with the function $f(x)$ we have described earlier. Let us consider $x_{\max} = 80$ and $x_{\min} = 0$ and let us take $n = 40$ points. We will have

$$\frac{\Delta x}{n} = 2. \tag{9.8}$$

In Figure 9.2 you can see the function $f(x)$ as continuous line, the crosses mark the points we sample in the grid search, and the black square mark the precise maximum of the function. The right plot shows a zoom around the maximum. You can see

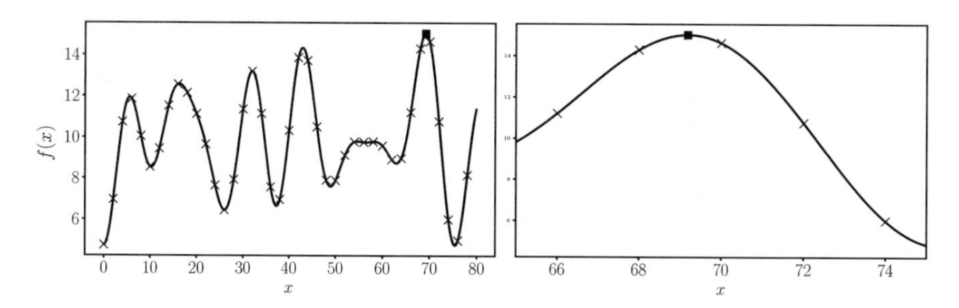

Fig. 9.2 the function $f(x)$ on the range $[0, 80]$. The crosses mark the point we sample in the grid search, and the black square mark the maximum.

how the points we sample in Figure 7-2 get close to the maximum, but don't get it exactly. Of course, sampling more points would get us closer to the maximum, but would cost us more evaluations of $f(x)$. This approach gives us $(\tilde{x}, \tilde{f}) = (70, 14.63)$ that is close to the actual maximum $(69.18, 15.03)$ but not quite right. Let us try the previous example varying the number of points (and therefore the interval Δx) we sample and let us see what results we get. We will vary the Δx from 1 to 20 in 0.5 steps. For each case we will find the maximum and its location as described earlier.

In Figure 9.3 we plot the distributions of the results. The black vertical line is the correct value of the maximum. As you can see the results vary quite a lot, and can be

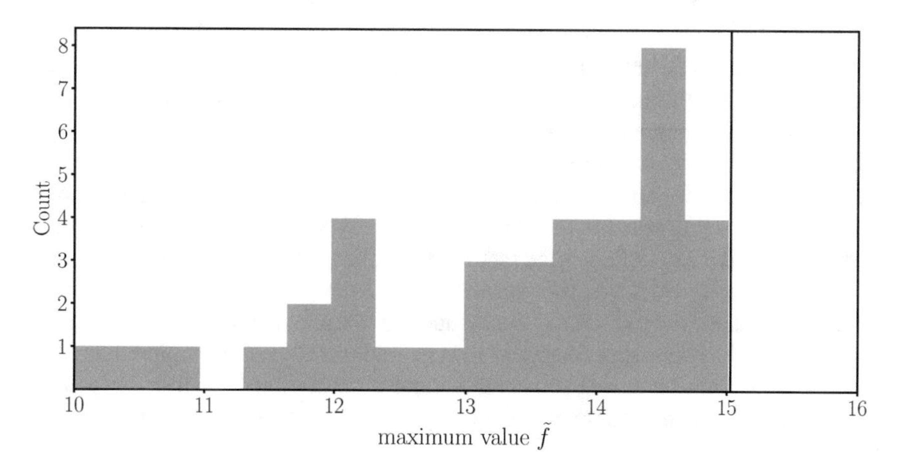

Fig. 9.3 The distribution of the results for $\tilde{f}$ obtained by varying the number of points n sampled in the grid search. The black vertical line indicates the real maximum of $f(x)$.

very far from the correct value. This tells us that using the wrong number of points can lead to very wrong results. As you can imagine, the best results are the ones with the smallest step Δx, since is more probable to get closer to the maximum. In Figure 9.4 you can see how the value of the found maximum varies with the step Δx.

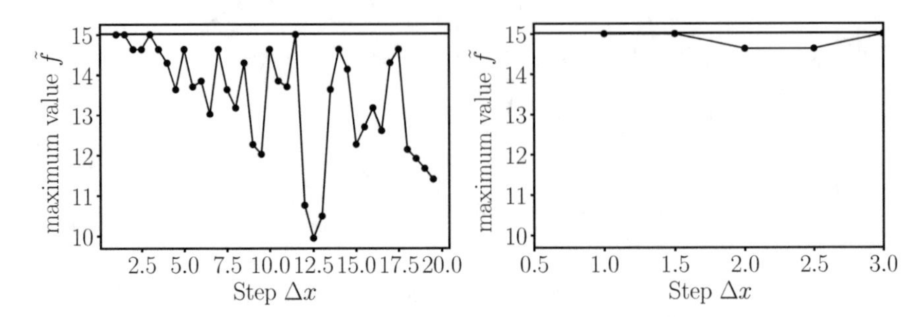

Fig. 9.4 The behaviour of the found value of the maximum vs. the x step Δx. The horizontal line slightly above 15 is the correct value of the maximum.

Warning Grid Search is Expensive

In the zoom in the right plot of Figure 9.4, is evident how smaller values of Δx give you better values of $\tilde{f}$. Note that a step of 1.0 means sampling 80 values of $f(x)$. If, for example, the evaluation takes one day, you will need to wait 80 days to get all the measurements you need. Grid search is a method that is efficient to be used only when the black-box function is cheap. To get good results, a large number of sampling points is usually needed.

To ensure that you are really getting the maximum, you should decrease the step Δx, or increase the number of sampling points, until the maximum value you find does not change appreciably anymore. In our example above, as we see from the right plot in Figure 9.4, we are sure that we are close to the maximum when our step Δx gets smaller than roughly 2.0, or, in other words, when the number of points sampled is greater or roughly equal to 40. Recall that 40 may seem quite a small number at first sight, but if $f(x)$ evaluates the metric of your deep learning model, and training takes, for example, two hours, you are looking at 3.3 days of computer time. Normally in the deep learning world 2 hours is not much for training a model, so make a quick calculation before starting a long grid search. Additionally, keep in mind that when doing hyper-parameter tuning you are moving in a multi-dimensional space (you are not optimising only one parameter, but many), so the number of evaluations needed get big very fast.

Let us make a quick example. Suppose that you decide you can afford 50 evaluations of your black box function. If you decide you want to try the following hyper-parameters

- Optimizer (RMSProp, Adam or plain GD) (3 values)
- Number of epochs (1000, 5000 or 10000) (3 values)

You are already looking at 9 evaluations. How many values of the learning rate can you then afford to try? Only 5! And with 5 values is not probable to get close to the optimal value. This example has the goal to let you understand how grid search is viable only for cheap black box functions. Remember that often, time is not the only problem. If you are using for example the google cloud platform to train your network, you are paying the hardware you use by the second. Maybe you have lots of time at your disposal, but costs may go over your budget very quickly.

9.4.2 Random Search

A strategy that is as "dumb" as grid search, but that works a lot better, is random search. Instead of sampling x points regularly in the range $(x_{\min}, x_{\max})$ you sample the points randomly. As we have done for grid search, you can see in Figure 9.5 the plot of $f(x)$, where the crosses mark the sampled points, and the black square the maximum. On the right plot, you see a zoom around the maximum.

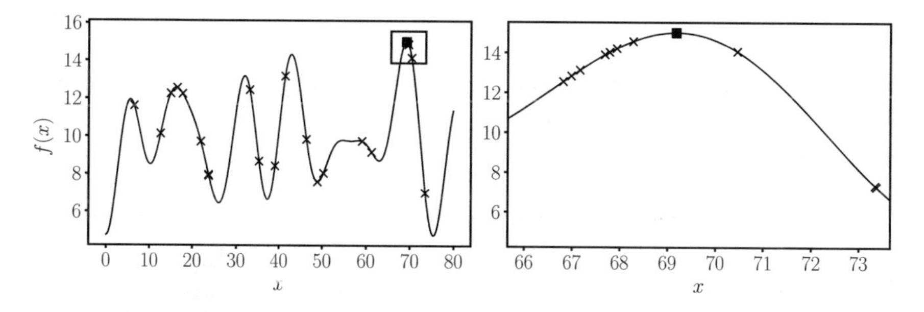

Fig. 9.5 the function $f(x)$ on the range $[0, 80]$. The crosses mark the point we sample with random search, and the black square mark the maximum.

In the right panel of Figure 9.5 you can clearly see how close we get to the maximum. The risk with this method is that, if you are unlucky, your random chosen points are nowhere close the real maximum. But that probability is quite low. Note that if you take a constant probability distribution for your random points you have the same probability of getting the points everywhere. Now it is interesting to see how this method performs. Let us consider 200 different random sets of 40 points, obtained by varying the random seed used in the code. The distributions of the maximum found $\tilde{f}$ is plotted in Figure 9.6.

Regardless of the random sets used, you get, in the most cases, very close to the real maximum. Figure 9.7 shows the distributions of the maximum found with random search varying the number of points sampled, from 10 to 80.

If you compare with grid search, random search is better at getting consistently results closer to the real maximum. Figure 9.8 shows a comparison between the

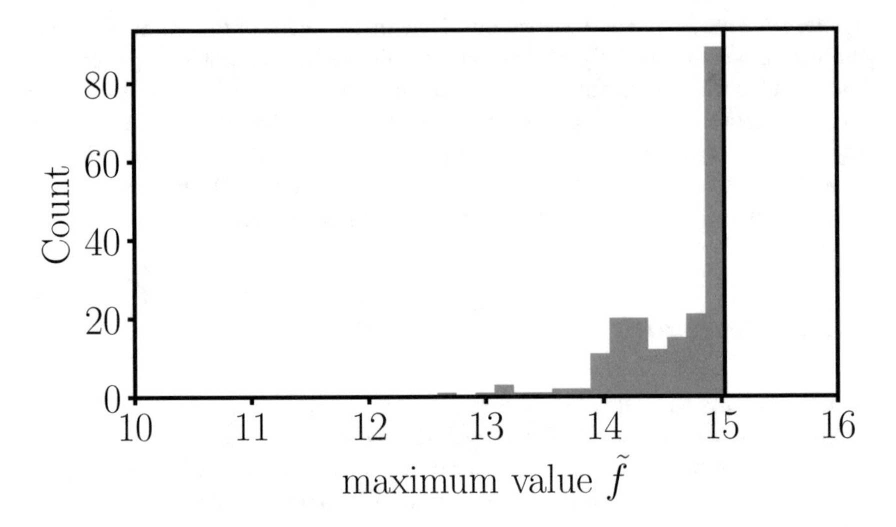

Fig. 9.6 distributions of the maximum found $\tilde{f}$ in 200 different random sets of 40 points selected with random search in the range $[0, 80]$.

distribution you get for your maximum $\tilde{f}$ when using the same number of sampling points n with random and grid search.

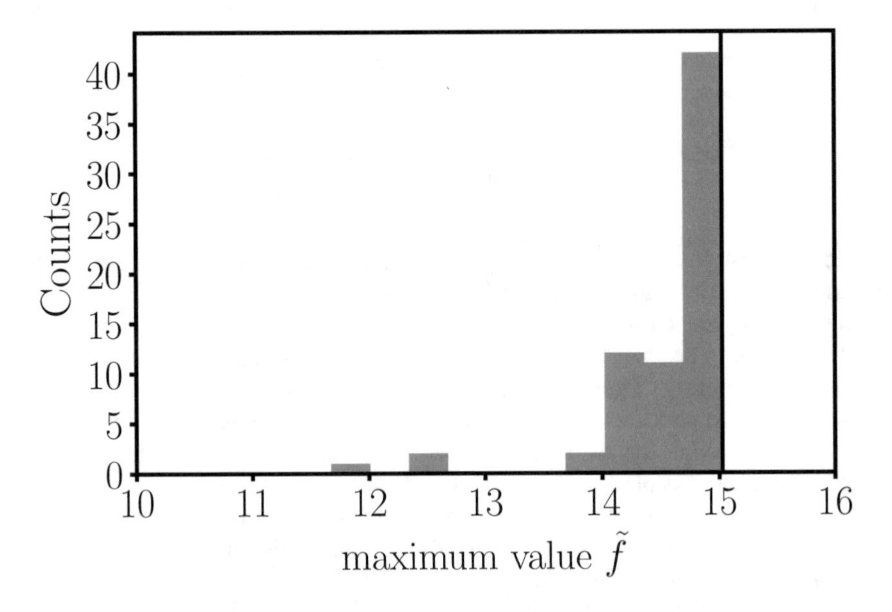

Fig. 9.7 The distribution of the results for $\tilde{f}$ obtained by 70 different random sets of 10 to 80 points sampled with the random search. The black vertical line indicates the real maximum of $f(x)$.

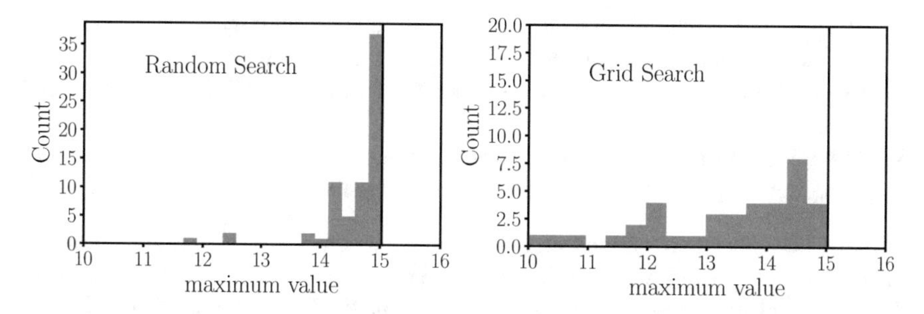

Fig. 9.8 The distribution of the results for $\tilde{f}$ obtained with random search (left panel) and with grid search (right panel) with the same number of points. You can see how random search outperforms grid search. The black vertical line indicates the real maximum of $f(x)$.

Is easy to see how on average random search is better than grid search. The values you get are consistently closer to the right maximum.

> ***Info*** **Random Search is More Efficient**
>
> Random search is consistently better than grid search, and you should use it every time possible. The difference between random search and grid search becomes even more marked when dealing with a multidimensional space for your variable x. hyper-parameter tuning is almost always a multidimensional optimisation problem.

If you are interested in a very good paper on how random search scales at high-dimensional problems, read the paper by J. Bergstra and Y. Bengio, Random Search for Hyper-Parameter Optimisation [2].

9.4.3 Coarse to Fine Optimisation

There is still an optimisation trick that helps with grid or random search. It is called "coarse to fine optimisation". Let us suppose that we want to find the maximum of $f(x)$ between $x_{\min}$ and $x_{\max}$. We will look at the idea of random search, but it works the same way with grid search. The following steps give you the algorithm you need to follow for this optimisation.

1. Do a random search in the region $R_1 = (x_{\min}, x_{\max})$. Let's indicate the maximum found with (x_1, f_1)
2. Consider now a smaller region around x_1, $R_2 = (x_1 - \delta x_1, x_1 + \delta x_1)$ for some δx_1 that we will discuss later and do again a random search in this region. Our hypothesis is, of course, that the real maximum lies in this region. We will indicate the maximum you find here with (x_2, f_2)

3. Repeat step 2 around x_2, in the region we will indicate with R_3 with a δx_2 smaller than δx_1 and indicate the maximum you find in this step with (x_3, f_3)
4. Now repeat step 2 again around x_3, in the region we will indicate with R_4 with a δx_3 smaller than δx_2
5. Go on in the same way as many times as you need until the maximum (x_i, f_i) in the region R_{i+1} does not change any more (at least in an appreciable way).

Usually, just one or two iterations are used, but theoretically, you could go on for many iterations. The problem with this method is that you cannot really be sure that your real maximum is truly in your regions R_i. But this optimisation has a big advantage if it does. Let us consider the case where we do a standard random search. If we want to have on average a distance between the sampled points of 1% of $x_{max} - x_{min}$ we would need around 100 points if we decided to perform only one random search, and consequently we would have to perform 100 evaluations. Now, let us consider the algorithm we just described. We could start with just 10 points in region $R_1 = (x_{min}, x_{max})$, here we will indicate the maximum we find with (x_1, f_1). Then let us take

$$2\delta x = \frac{x_{max} - x_{min}}{10} \tag{9.9}$$

and let us take again 10 points in region $R_2 = (x_1 - \delta x, x_1 + \delta x)$. In the interval $(x_1 - \delta x, x_1 + \delta x)$ we will have on average a distance between the points of 1% of $x_{max} - x_{min}$, but we just sampled our functions only 20 times instead of 100. For example, let us just sample our function we used above between $x_{min} = 0$ and $x_{max} = 80$ with 10 points with random search. This gives us the maximum location and value of $x_1 = 69.65$ and $f_1 = 14.89$, so not bad but not yet as precise as the real values 69.18 and 15.027. Now let us sample again 10 points around the maximum we have found in the regions $R_2 = (x_1 - 4, x_1 + 4)$. This will give us values of $x_2 = 69.19$ and $f_2 = 15.027$. The results now are much close to the correct value, and with only 20 evaluations.

Warning **Dangers of Coarse to Fine Optimisation**

If you decide to use this method, keep in mind that you will still need a good number of points at the beginning to get close to the maximum before refining your search. After you are relatively sure to have points around your maximum, you can use this technique to refine your search.
If the maximum is far from the one found in the first iteration, you will end up only with a local maximum or worse with just a random point that is far away from your maximum. Unfortunately, especially with neural networks, is almost impossible to find out if you are in such a case. Thus testing with various number of points may be necessary.

9.4.4 Sampling on a Logarithmic Scale

Now there is a last small subtlety that we need to discuss. Sometimes you will find yourself in a situation where you want to try a wide range of possible values for a parameter, but you know from experience that probably the best value of it is in a specific range. Let us suppose that you want to find the best value for the learning rate for your model and decide to test values from 10^{-4} to 1, but you know, or at least expect, that your best value lies probably between 10^{-3} and 10^{-4}. Now let us suppose that you are working with grid search, and let us suppose that you sample 1000 points. You may think you have enough points, but you will get

- 0 point between 10^{-4} and 10^{-3}
- 8 points between 10^{-3} and 10^{-2}
- 89 points between 10^{-1} and 10^{-2}
- 899 points between 1 and 10^{-1}

You probably want to sample in a much finer way for smaller values of the learning rate than for bigger one. What you should do is to sample your points on a logarithmic scale. The basic idea is that you want to sample the same number of points between 10^{-4} and 10^{-3}, 10^{-3} and 10^{-2}, 10^{-1} and 10^{-2} and 10^{-1} and 1. To do that, you can use the following algorithm.

- First select a random number r between 0 and minus the absolute value of the highest number of the power of 10 you have, in this case -4.
- then your points for the grid search can be generated by using 10^r.
- You repet the algorithm for as many points you need.

In our example, by following this algorithm, you would get 250 points in each region, as you can easily check. Now you can see how you have the same number of points between the different powers of 10. With this simple trick you can ensure that you get enough points also in region of your preferred range, where otherwise you would get almost no points. Remember that in this example, with 1000 points, with the standard method we get zero points between 10^{-3} and 10^{-4}. This range is the most interesting for the learning rate, so you want to have enough points in this range to optimise your model. Note that the same applies to random search. It works in the exact same way.

9.5 ■ Overview of Approaches for Hyper-parameter Tuning

In this chapter we have scratched the surface on what is hyper-parameter tuning and how to perform it. It is useful to see how many variations and approaches are available in addition to those discussed here. The following list will give the reader an overview of the most known approaches for various use cases and some interesting references.

1. **Grid Search:** An exhaustive search over a specified parameter space [3] (discussed in this chapter)
2. **Random Search:** Randomly selecting hyper-parameter combinations over the parameter space (discussed in this chapter).
3. **Bayesian Optimization:** Uses a probabilistic model to select hyper-parameters to evaluate based on past results (see for example [4], a seminal work by Snoek *et al.* in applying Bayesian Optimization to hyper-parameter tuning in machine learning algorithms, [5], a review paper by Shahriari *et al.* that provides a comprehensive overview of Bayesian Optimization, its developments, and applications, the book by Rasmussen and Williams [6], to understand Gaussian Processes, a key component in many Bayesian Optimization algorithms or the book by yours truly [7] for a practical implementation in Python).
4. **Gradient-Based Optimisation:** Suitable for differentiable hyper-parameters, adjusting them via gradient information (see for example [8]).
5. **Evolutionary Algorithms:** Utilizes algorithms like Genetic Algorithms to simulate natural evolution in hyper-parameter optimization (see for example [9] by Eiben *et al.*, the book by Goldberg [10] that provides foundational knowledge about genetic algorithms, covering their application in search, optimization, and machine learning, or the very interesting article by Zhang *et al.* [11] *et al.*, that focuses on the application of evolutionary algorithms, specifically for evolving the architecture of deep neural networks, an aspect of hyper-parameter optimization.).
6. **Hyperband:** A bandit-based approach that efficiently allocates computational resources (see for example [12]).

References

1. Prajit Ramachandran, Barret Zoph, and Quoc V. Le. Searching for activation functions. *arXiv:Computation and Language*, 2017.
2. James Bergstra, Remi Bardenet, Yoshua Bengio, and Balazs Kegl. Algorithms for hyper-parameter optimization. In *Advances in Neural Information Processing Systems*, pages 2546–2554, 2011.
3. James Bergstra and Yoshua Bengio. Random search for hyper-parameter optimization. *Journal of Machine Learning Research*, 13:281–305, 2012.
4. Jasper Snoek, Hugo Larochelle, and Ryan P. Adams. Practical bayesian optimization of machine learning algorithms. *Advances in Neural Information Processing Systems*, 25, 2012.
5. Bobak Shahriari, Kevin Swersky, Ziyu Wang, Ryan P. Adams, and Nando de Freitas. Taking the human out of the loop: A review of bayesian optimization. *Proceedings of the IEEE*, 104(1):148–175, 2016.
6. Carl Edward Rasmussen and Christopher K. I. Williams. *Gaussian Processes for Machine Learning*. The MIT Press, 2006.
7. Umberto Michelucci. *Applied Deep Learning with TensorFlow 2: Learn to Implement Advanced Deep Learning Techniques with Python*. Amazon Digital Services, 2022.
8. Dougal Maclaurin, David Duvenaud, and Ryan P. Adams. Gradient-based hyperparameter optimization through reversible learning. In *International Conference on Machine Learning*, pages 2113–2122, 2015.
9. Agoston Eiben and J.E. Smith. *Introduction to Evolutionary Computing*. Springer, 2015.

10. David E. Goldberg. *Genetic Algorithms in Search, Optimization, and Machine Learning*. Addison-Wesley, 1989.
11. Xingyi Zhang, Jeff Clune, and Kenneth O. Stanley. Evolving deep neural networks. In *Artificial Life Conference Proceedings*, pages 258–265, 2018.
12. Lisha Li, Kevin Jamieson, Giulia DeSalvo, Afshin Rostamizadeh, and Ameet Talwalkar. Hyperband: A novel bandit-based approach to hyperparameter optimization. *Journal of Machine Learning Research*, 18(185):1–52, 2017.

Chapter 10
Feature Importance and Selection

I much prefer being occasionally imprecise but understandable to being completely accurate but incomprehensible.
George F. Simmons

This chapter offers an in-depth exploration of various methods used to assess feature importance in machine learning models. Initially, it highlights the importance of identifying key features in a machine learning model, using examples from the healthcare and finance sectors to illustrate its significance. The chapter categorises feature importance assessment methods into three types: filter, wrapper, and embedded methods. Filter methods are preprocessing steps that select features independently of the model, with examples that include the variance threshold, the Chi-square test, the information gain, and the correlation coefficient. These methods are notable for their computational efficiency and scalability. The section on wrapper methods delves into techniques like recursive feature elimination, forward feature selection, backward feature elimination, exhaustive feature selection, and information content elimination. These methods are flexible and consider interactions between features, which can lead to better model performance but at the cost of higher computational intensity. Embedded methods, such as LASSO, ridge regression, elastic net, and decision trees, are integrated into the model training process and specific to each model, making them somewhat challenging to interpret in scientific contexts. The chapter also provides a detailed explanation of forward feature selection, backward feature elimination, information content elimination, and permutation feature importance.

10.1 Introduction

In machine learning, not everything revolves around training models to classify or predict something. Once a model has been trained, under the assumption that its performance is good enough, a natural question to ask is which of the features used as input is the most important. This information is quite important, as it can tell us something about the phenomena we are trying to model with machine learning. Let us look at two examples from different domains to illustrate this.

In the healthcare sector, consider a machine learning model designed to predict hospital readmissions. This model might take into account a variety of patient data,

U. Michelucci, *Fundamental Mathematical Concepts for Machine Learning in Science*,
https://doi.org/10.1007/978-3-031-56431-4_10

such as age, diagnosis, lab test results, length of stay, and medication history. The role of feature importance here is quite important. First, it helps healthcare providers pinpoint the most critical predictors of readmission. For example, if the model identifies the length of the previous hospital stay and specific lab test results as the top predictors, hospitals can focus on optimising these aspects to reduce readmission rates. Furthermore, understanding which features significantly influence readmissions allows more efficient resource allocation and the development of targeted intervention strategies. If, for example, the model reveals that certain medications are the leading predictors, hospitals can examine and refine their prescription practices. Furthermore, this analysis can inform policy making and preventive measures. If socio-economic factors such as income level or access to care are found to significantly impact readmissions, it could lead to the creation of community health programmes that address these issues.

In the financial sector, consider a credit scoring model that banks use to determine the creditworthiness of loan applicants. This model may include features such as income, employment history, credit history, age, and marital status. The importance of understanding the significance of features in this context is crucial for several reasons. It plays a key role in risk management by helping banks identify which features most strongly predict credit risk. If the model shows that credit and employment history are important, banks can give these factors more weight in their credit decision processes. Additionally, assessing the importance of features is essential to ensure fair lending practices. It can reveal whether noncredit-related factors, such as marital status or age, are disproportionately influencing credit decisions, which is vital for compliance with regulatory standards and ethical lending. Furthermore, recognising the most impactful features enables financial institutions to better customise their financial products. If income level is a significant predictor, for instance, banks might develop different credit offerings for different income brackets.

Therefore, assessing the importance of features is often a fundamental step in any machine learning project. Different models, by design, can give an estimate of the relevance of the input features out of the box. But in this chapter, we discuss only a few approaches that can be used regardless of the machine learning model used. This makes them applicable in a wide variety of use cases and, given their design, allows a profound interpretation of the results that can be easily translated into a language that the end user can understand and profit from. Let us first start with a taxonomy of feature-important and selection methods.

10.2 Feature Importance Taxonomy

To assess feature importance, there are three types of approaches.

1. **Filter methods**: filter methods select a subset of features as a pre-processing step, independently of the chosen model. For example, you could only choose features that have a variance greater than a certain threshold.

2. **Wrapper methods**: wrapper methods use machine learning models as black-boxes to score subset of features. Examples are forward, backward or exhaustive feature selection.
3. **Embedded methods**: embedded methods perform variable selection in the model training process and are usually specific to a given machine learning model.

In the following subsections, some notable examples of the previously mentioned approaches are listed to give the reader more information. After that, two wrapper methods will be discussed at length since they are widely applicable and general since they do not depend on any chosen model.

10.2.1 ■ Filter Methods

It is useful to briefly discuss some filter methods to give the reader some idea about the existing approaches.

- **Variance Threshold**: this method removes all features whose variance is lower than some threshold. It is useful for removing features that are constant or almost constant, and therefore do not provide much information for the training of the model.
- **Information Gain**: commonly used in decision trees, information gain measures the reduction in entropy or surprise[1] from transforming a dataset in some way. It is used to decide which feature to split on at each step in building a tree (see, for example, [1]).
- **Correlation Coefficient**: this is a statistical measure that expresses the extent to which two variables are linearly related. Features that are highly correlated with the target variable are considered important. For example, experiments with a low correlation with the target variable can then be neglected.

These methods are typically easy to implement and computationally less expensive than others, such as wrapper or embedded methods, and scale well to very large datasets. They are independent of the chosen machine learning model and can be used as a *pre-processing* step. Their main advantages are that they help eliminate redundant or irrelevant features based on a statistical analysis, and thus may help reduce overfitting and do a kind of dimensionality reduction.

However, they do not consider how a specific model may use the features, so they may remove features that could be helpful for a specific model. Additionally, they consider each feature independent from each other, and thus simply ignore feature correlations, which in some cases may help to get good prediction performance.

To summarise, filter methods work best for an initial feature reduction; however, if prediction performance is important other approaches as *wrapper* or *embedded* are more advisable.

[1] In information theory, *surprise* is a concept that is closely related to the idea of entropy and information content. It essentially refers to the unexpected or unpredictable nature of a piece of information. The more surprising or less predictable an event is, the more information it carries.

10.2.2 ■ Wrapper Methods

Wrapper methods assess subsets of features to determine their effectiveness by using a model as a black-box. They use a model that is trained on a combination of features and assign a score based on some chosen metric (for example accuracy or MSE) to the feature set. The most well-known and used wrapper methods are the following.

- **Recursive Feature Elimination (RFE)**: RFE works by recursively removing features and building a model on the remaining features. It uses a metric to identify which features (or combinations of features) contribute the most to predicting the target variable. RFE will repeatedly construct a model and choose the best or worst performing feature, setting it aside, and then repeat the process with the rest of the features. In Python is particularly easy to implement; check (as usual) the official documentation for details at `https://scikit-learn.org/stable/modules/generated/sklearn.feature_selection.RFE.html`.
- **Forward Feature Selection**: This is an iterative method in which in each iteration, one keeps adding the feature which improves our model the most (according to some metric) until an addition of a new variable does not improve the performance of the model anymore (more details are given in the following sections) or some other condition is met.
- **Backward Feature Elimination**: In this method, one starts with all the features and removes the least significant feature at each iteration (the one that has the least impact on the model's performance). We repeat this until no improvement is observed on the removal of features (more details are given in the following sections) or some other condition is met.
- **Exhaustive Feature Selection**: This method examines all possible combinations of features and selects the combination that yields the best model performance. It is computationally quite intensive as the number of possible combinations can be very large, especially with a large number of features. Thus, this method is rarely used.
- ***Y*-Randomisation**: this method is a technique used to assess the importance of features within a dataset by randomly shuffling the values of each feature individually and observing the impact on model performance, thereby helping to validate the predictive reliability of the model.
- ***Y*-Elimination**: this easy-to-nderstand method is an approach used to assess the importance of features by setting to zero (or some other constant) the values of each feature individually and observing the impact on model performance.

Wrapper methods have several advantages. They take into account correlations between features, for example. In fact, unlike filter methods, which evaluate each feature independently, wrapper methods consider the dependency and interaction between features. This may lead to better performing models. They are quite flexible, as they can be used with any machine learning algorithm, making them usable in almost any case.

Info **Subsets of features for science**

Wrapper methods are particularly useful in scientific contexts, as they ensure that the chosen features are directly related to a specific research question. Take, for example, a classification task in which the selected feature subset is tailored for that classification. If the task shifts to regression, the focus also shifts to selecting features pertinent to predicting the continuous variable in the regression problem.

On the other hand, wrapper methods can be computationally expensive, especially for datasets with many features, since one typically needs to train a model several times. Being usable with any model is, as mentioned, an advantage, but can also be a disadvantage. In fact, since the feature selection process is dependent on the model used, the selected features might be optimal for one type of model but not for another, limiting the generalisation power of the feature selection process. This is normally not an issue, since the importance of features is **always** related to the problem you are trying to solve. Recall that something is only important for the sake of something and not in a vacuum.

10.2.3 ■ Embedded Methods

Embedded methods integrate the feature selection process in the learning algorithm. Since each model will use a different approach to determine the importance of the features, difficulties arise because each model adopts a unique approach to assess feature importance, making it hard and often nearly impossible to connect the identified importance back to the intrinsic characteristics of the features themselves. This makes them slightly less useful in scientific settings. Nevertheless, it is useful to have some idea about the most well-known. Here are four of the most important embedded methods.

- **LASSO** (Least Absolute Shrinkage and Selection Operator): LASSO is a regression analysis method that performs both variable selection and regularisation[2] at the same time. It adds a penalty equal to the absolute value of the magnitude of the coefficients (essentially the same as the ℓ_1 regularisation techniques known in neural networks) to the function that needs to be minimised. This approach effectively shrinks some of the model weights to zero, thus performing, although in a different way from the previously described methods, some indirect feature selection. LASSO is particularly useful when dealing with datasets with many features, as it promotes a simpler and more interpretable model with fewer predictors. A very good introduction to the method can be found in the wonderful book *an Introduction to Statistical Learning* by James, Witten, Hastie and Tibshirani

[2] Regularisation is a technique in machine learning that reduces overfitting by penalising complex models, thereby enhancing the model's generalisation ability on unseen data.

[2]. Note that it is not always possible to extract which feature is irrelevant and which is not with this method, so interpretability suffers.

- **Ridge Regression**: while ridge regression is primarily known for its regularisation capability, it is also used in feature selection. It introduces a penalty term that is the squared magnitude of the coefficients (essentially the same as the ℓ_2 regularisation techniques known in neural networks) and have a similar effect then the LASSO regularisation.
- **Elastic Net:** Elastic net is a hybrid approach that combines LASSO and Ridge regression penalties. It is useful for selecting features in situations where there are multiple correlated features. Elastic Net aims to maintain the grouping effect like Ridge and the selection ability of LASSO. It is particularly effective when dealing with highly dimensional data.
- **Decision Trees** (and algorithms derived such as Random Forest, Gradient Boosting Machines): decision trees inherently perform feature selection by selecting the most informative features to split the data at each node. Algorithms like Random Forest and Gradient Boosting Machines, which are ensembles of decision trees, also include feature importance scores as part of their output, which can be used to identify key features.

Warning **Embedded Methods in Science**

Embedded methods may seem a comfortable way to perform feature importance assessment, as models like decision trees automatically give an *importance score* to each feature. The problem is that it is almost impossible to explain the meaning of this score in a way that could be related to the science behind the data. Additionally, often simply by training with different datasets, the order of feature may (and often will) change, and thus using this information to extract an explanation of the correlation between inputs and outputs is almost impossible.

10.3 Forward Feature Selection

Now let us describe the first wrapper method that we discuss in this book: forward feature selection (FFS). It is easy to implement and easy to interpret.

Let us suppose that we have a dataset with n features that we will indicate with F_i with $i = 1, \cdots, n$. F_i can be, for example, the age of a patient, his or her weight, the concentration of a chemical component, etc. To simplify the discussion, let us indicate a hypothetical model trained with a set of features $\{F_1, ..., F_n\}$ with $\mathcal{M}(\{F_1, ..., F_n\})$. Forward Feature Selection (FFS) works according to the steps below. note that we will describe the algorithm by using a hold-out approach. In principle, a better approach would be to use a cross-validation approach (for example, a k-fold cross validation). We will first discuss the *easier* version with a hold-out approach and then discuss the most complex case. Here are the first steps.

1. The dataset D is splitted in a training D_T and validation D_V portions.
2. n models are trained on D_T, each on a single feature F_i: $\mathcal{M}(\{F_i\})$, with $i = 1, ..., n$.
3. The feature that give the best model $\mathcal{M}(\{F_i\})$ performance on D_V (according to a specified criterion, usually a performance metric such as accuracy for classification or R-squared for regression) is chosen. Let us indicate this first feature as $\mathcal{F}_1$. Note that naturally $\mathcal{F}_1$ will be one of the features contained in the set $\{F_1, ..., F_n\}$ or, in other words, $\mathcal{F}_1 = F_j$ for some j.
4. $n - 1$ models are trained on D_T, each using two features $\mathcal{F}_1$ and F_i: $\mathcal{M}(\{\mathcal{F}_1, F_i\})$ for $i \neq j$.
5. The feature k that gives the best performance on D_V of the models trained in Step 4 is then chosen and indicated with $\mathcal{F}_2 = F_k$.

This process goes on adding at each iteration one feature. At the end of the process, we will have an ordered set of features $\mathcal{F}_1, .., \mathcal{F}_n$. Their order will indicate the most important one (in the first place) and the least important one (in the last place) sorted according to the algorithm specified above.

To summarise, the FFS process starts with an empty model and iteratively adds features to the model. At each step, it evaluates all the features that are not yet in the model and adds the one that most improves the model according to a specified criterion, usually a performance metric like accuracy for classification or R-squared for regression.

If we are only interested in getting the **order** of the features based on their importance, we are done. If we are doing this process to select the top (say) 50% of the features, then we will need a third dataset. To make the idea easy to follow, let us imagine a hypothetical case. Suppose we are trying to classify some kind of inputs, and that we have 10 features at our disposal. From the problem we might know, for example, that the 10 features are surely too much, and we expect that 4 or 5 should be enough. We decide then to use FFS. Suppose that we find that $\mathcal{M}(\{F_3\})$ gives us an accuracy of 78%. Then we start adding a second feature and find that $\mathcal{M}(\{F_3, F_5\})$ gives an accuracy of 82%. Then we find that $\mathcal{M}(\{F_3, F_5, F_1\})$ gives us an accuracy of 87%. Suppose now that adding a fourth feature would give us an increase in accuracy of only 0.5%. We may decide that 87% is enough, because maybe the features F_3, F_5 and F_1 are easy to measure and thus easy to use in the field in a real-life scenario. At this point, we can stop adding new features. Now we can train our model on $D_T \cup D_V$. We still need a third dataset to validate this model, since our feature selection could work well **only** for the validation dataset, thus we need to test it on new data. We now test our newly trained model on D_{test} to check its performance. We have, of course, to check, as described in the chapter on model validation (Chapter 7) whether we are overfitting or not as usual. This is, with an example, the approach you should use if you want to use FFS for feature number reduction. To have three datasets, typically one splits the original dataset in three parts, for example in 60%/20%/20% portions (but it could be 80%/10%/10%). Let us discuss an example to highlight some issues that you may find when using FFS.

10.3.1 Forward Feature Selection Practical Example

To give you an idea about how this method works, we will consider the breast cancer dataset (see the info box below for a description) and test FFS on it.

Info The Breast Cancer Dataset

The Breast Cancer Wisconsin (Diagnostic) dataset [3] is a renowned dataset in the field of medical machine learning, especially for binary classification tasks. Hosted by the UCI Machine Learning Repository and integrated into scikit-learn, this dataset offers a comprehensive set of features derived from fine needle aspirate (FNA) images of breast masses. It is instrumental in developing and evaluating machine learning models for cancer diagnosis.
The dataset comprises 569 instances, each representing a case with a set of features calculated from a digitised image of a FNA of a breast mass and an associated diagnostic label (malignant or benign). The features represent characteristics of the cell nuclei present in the image.
Detailed Feature Description:

- Each instance in the dataset is characterised by 30 features. These features are computed from ten primary measurements taken for each cell nucleus in the image:

  1. **Radius**: mean of distances from the center to points on the perimeter.
  2. **Texture**: standard deviation of the greyscale values in the image.
  3. **Perimeter**: the size of the boundary of the nucleus.
  4. **Area**: The area of the nucleus.
  5. **Smoothness**: local variation in the lengths of the radius of the nucleus.
  6. **Compactness**: calculated as $\frac{\text{perimeter}^2}{\text{area}-1.0}$, this measures how compact the nucleus is.
  7. **Concavity**: the severity of concave portions on the nucleus contour.
  8. **Concave Points**: the number of concave portions on the nucleus contour.
  9. **Symmetry**: symmetry of the shape of the nucleus.
  10. **Fractal Dimension**: "coastline approximation" quantifies the complexity of the nucleus shape, calculated as $1 - \frac{\text{coastline}}{\text{area}}$.

- For each of these primary measurements, three additional metrics are computed:

  1. **Mean**: average of the measurement across all nuclei in the image.
  2. **Standard Error**: the standard error of the measurement.
  3. **Worst**: the mean of the three largest measurements, indicating the presence of extreme outliers or anomalies.

The target variable of the dataset is binary, indicating whether breast cancer is *malignant* or *benign*.

As a model, we will use a Naïve Bayes classifier (the model itself is not relevant for this discussion). I have split the data set D into two portions: 80% for training (D_T) and 20% for validation (D_V). The best feature at each iteration has been determined with a five-fold cross validation on D_T. In Figure 10.1 you can see the accuracy plotted as a function of the number of features (I stopped at 10 in this example) found by the FFS algorithm. There are several important things to note about this result. First, the behaviour is generally what we expect. The accuracy starts (already

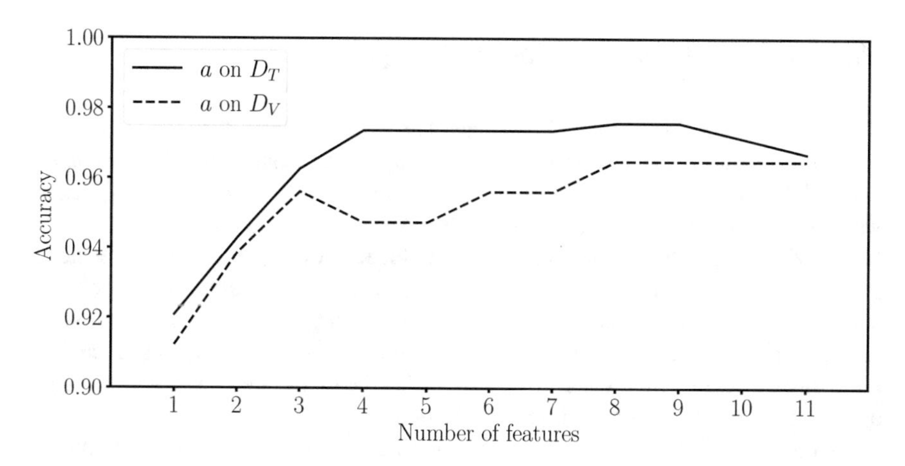

Fig. 10.1 The accuracy of a Gaussian Naïve Bayes classifier trained on the breast cancer dataset [3]. The best feature at each iteration has been determined with a 5-fold cross validation on the training dataset D_T(obtained by random sampling 80% of the data from the entire dataset).

from a fairly high value of slightly below 92%) from a *low* value and increases when we add more features. This is understandable since by adding features, we expect to add information that the model can use to make better predictions. What is interesting is that the accuracy is not monotone and oscillates. The accuracy on D_T, for example, when using 10 features, is lower than when using 9. This is something that can happen quite often when using FFS. This result highlights the fact that adding features may make the predictions of a model worse. It can happen that the feature actually confuses the model. Looking at the plot in Figure 10.1 one can then choose the best number of features (of course, when using the FFS algorithm it is possible to get *which* features are used in what combination) according to specific criteria. For example, when using all features the model reaches an accuracy of 94.7% and 91.2% on the D_T and D_V respectively (in the interest of fairness it is showing a bit of overfitting, a little detail we will conveniently overlook here). Using the best 8 features it reaches an accuracy of 97.6% and 96.4% on D_T and D_V, respectively. Showing that sometimes using fewer features might be a way to get better performance!

Warning **Less Features, Better Performance?**

Sometime by using less features your model can perform better. Not all features are created equal, and sometimes they may contain errors or simply give incorrect information about a particular input. Doing FFS is a very viable way to check which combination of features gives you the best performance. It is worth checking.

10.4 Backward Feature Elimination

Backward Feature Elimination (BFE) takes a different approach than forward feature selection (FFS). Here is how it works: We start by dividing our dataset into two parts, D_T and D_V. Initially, we train a model using all the features, $\mathcal{M}(F_1, ..., F_n)$. Then, we enter a process of elimination. In each round, we remove one feature from the full set $F_1, ..., F_n$, train the model with the reduced set of features, and evaluate its performance (using metrics such as accuracy or MSE). The feature whose removal least negatively impacts the model's performance, indicating its not so significant role in predictions, is identified. This methodical elimination continues, each time removing the feature that, when absent, degrades the model's effectiveness the least, until we are left with the most impactful features.

In Figure 10.2 you can see the accuracy behaviour while using BFE on the same dataset (namely the breast cancer dataset) described in the previous section. Overfitting pops up again, but we will simply ignore it in this discussion, as we have done in the FFS discussion.

Warning **FFS or BFE**

As shown in the example discussed above, the two methods produce very different results. While the FFS helps to get a good model with high accuracy, the BFE falls a bit short in this regard. As usually in machine learning, there is no golden rule that works every time, and testing is strongly recommended. Here are the main differences between them.

- **Starting Point and Direction:** FFS begins with an empty model and progressively adds features. It starts with no features and, at each iteration, evaluates and adds the feature that improves the model's performance the most, until no significant improvement is observed, or some other criteria is met. BFE, on the other hand, starts with a full model that includes all available features. It then progressively removes the least significant

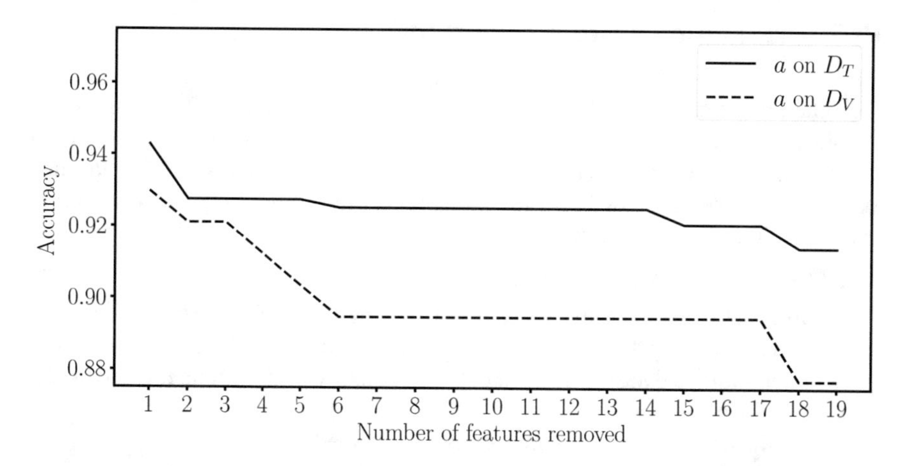

Fig. 10.2 The accuracy of a Gaussian Naïve Bayes classifier trained on the breast cancer dataset [3]. The worst feature to eliminate at each iteration has been determined with a 5-fold cross validation on the training dataset D_T(obtained by random sampling 80% of the data from the entire dataset).

feature, one at a time, that least deteriorates the model's performance, until a desired number of features or performance criterion is met.

- **Computational Complexity:** FFS can be less computationally intensive, especially when the initial number of features is large (since you start with one feature, then two, and so on). Since it begins with no features and adds them incrementally, it can stop early when sufficient performance is achieved, potentially evaluating fewer total feature combinations. BFE can be more computationally demanding in the case of large feature sets, as it starts with the full model and has to evaluate the impact of removing each feature iteratively, which can be time-consuming and resource-intensive.
- **Sensitivity to Redundant Features:** FFS can be more sensitive to redundant features. Since it adds features based on immediate improvement, it may select redundant features if they show an initial improvement, even if they do not contribute significantly when combined with later-added features. BFE is less prone to retain redundant features. Since it evaluates the model's performance with all features initially and then removes them one by one, it has a better chance of identifying and eliminating features that, while individually significant, may be redundant in the presence of other features.

There is no clear *better* method for feature selection or importance assessment and testing and knowledge about the problem must be used.

10.5 Permutation Feature Importance

Permutation feature importance (PFI) uses a different approach. The first step is to split the dataset in training D_T and validation D_V sets (as usual in a hold-out approach). Firstly, a model is trained on D_T and its performance (according to some metric, such as accuracy or MSE, for example) is evaluated on D_V. This will serve as our baseline. The process will continue as follows.

- Select one feature from your dataset.
- Randomly shuffle the values of this feature across all samples in your D_V dataset, effectively breaking the relationship between this feature and the target variable.
- With this shuffled dataset, reevaluate its performance on D_V. Note that the model is **not** trained again.
- Record the change in performance compared to the baseline.
- Restore the feature to its original order before proceeding to the next feature.

The importance of each feature is then assessed from the way the performance metric changes when that feature's values are shuffled. A greater decrease in performance is associated with a higher importance of the feature. To obtain a sorted list of features, you can rank them in terms of their importance to the predictions of the model, as discussed.

This approach may seem a reasonable approach that is very practical, as no multiple model trainings are necessary. It is surely computationally efficient. Regrettably, there are certain less favourable aspects associated with this approach. Here are the main ones.

- Since the method involves random shuffling of feature values, the results can sometimes be inconsistent, especially with smaller datasets. This randomness can lead to a large variability in importance scores on different runs. This makes it difficult to interpret the importance of features.
- PFI can struggle with datasets where features are highly correlated. When a feature is shuffled, the model might still make accurate predictions using the correlated features, leading to an underestimate of the importance of the shuffled feature. In this case, for example, eliminating redundant features before using PFI may be a solution to the problem.
- As the model is not retrained after shuffling each feature, this method does not account for the model's ability to adapt. In real-world scenarios, especially with flexible models like neural networks, the model might adjust to the absence or alteration of a feature if it is retrained.
- Importance scores are specific to the model and its current state of training. They do not necessarily represent universal or intrinsic importance of the features and can vary for different models or different instances of training the same type of model. In science, this presents a significant drawback of the method, as scientists seek to understand the importance of features irrespective of the model selected.
- If not interpreted carefully, the results can be misleading. For instance, a feature might seem unimportant not because it is irrelevant but because the model failed

to learn its importance properly during training. Therefore, this would be a model limitation and not an intrinsic characteristic of the data.

10.6 Information Content Elimination

Information Content Elimination[3] is simpler than Permutation Feature Importance. This method involves removing all information from a specific feature by setting its values to zero (or another constant). After training a model with all features, you modify the dataset by setting values of one feature to zero, and then assess the model's performance on this altered dataset without retraining. This process is repeated for each feature individually. The feature whose removal causes the most significant decrease in model performance is deemed the most critical. The criticisms listed in the previous section apply also to this method.

10.7 Summary

In general, it is important to understand that the **importance** of a feature must be defined in the context of the problem to be solved. For example, for a researcher, a feature may be **not** important if does not show enough variance (so it is predominantly constant), for another it could be a feature that, when removed, does not affect the prediction of a given model type. According to **your** definition, you should choose one of the previously discussed methods that matches your understanding of what a good or bad feature is. Feature importance does not live in a vacuum but is **always** related to the specific problem you are trying to solve.

10.8 ■ SHapley Additive exPlanations (SHAP)

This chapter, while surely not exhaustive in its coverage, would be notably more incomplete without at least a short discussion on SHAP values. The topic is too complex to discuss it at length in this book, and I will only give very short explanations and point the reader to an important reference to learn more about the method.

SHAP (SHapley Additive exPlanations) values are a tool used in machine learning to explain how different features in a model contribute to each individual prediction. Imagine that you are trying to predict house prices, and your model uses features such as location, size, and age of the house. SHAP values work by uncovering how much each feature (like location or size) contributes to the final prediction (the

[3] The name comes from me, as I was not able to find a name in the literature. So if you look for it you will not find it, at least under this name.

price) for each specific house. They do this by comparing what the model predicts with each included feature versus what it would predict if that feature was absent, effectively measuring the impact of each feature. The uniqueness of SHAP values lies in their fairness, rooted in game theory. If you think of each feature as a player on a team, the SHAP values calculate how to fairly distribute the credit for a win (the prediction) among the players (the features), based on their individual contributions. For example, if a house in a prime location (a key feature) is predicted to have a high price, the SHAP value will show a significant positive contribution from the location. On the other hand, if the house is old, that might negatively impact the price, and this will be reflected in the SHAP value for the house's age.

The mathematics behind the method is not easy, and the interested reader can consult the original paper [4] where the complete mathematics explanation is discussed at length.

References

1. J. Ross Quinlan. *C4.5: Programs for Machine Learning*. Morgan Kaufmann, San Mateo, CA, 1993.
2. Gareth James, Daniela Witten, Trevor Hastie, and Robert Tibshirani. *An Introduction to Statistical Learning*. Springer New York, 2013.
3. William H Wolberg, W Nick Street, and Olvi L Mangasarian. Breast cancer wisconsin (diagnostic) data set. *Uci machine learning repository*, 1992.
4. Scott M. Lundberg and Su-In Lee. A unified approach to interpreting model predictions. In *Advances in Neural Information Processing Systems*, volume 30, pages 4765–4774. Curran Associates, Inc., 2017.

Index

U. Michelucci, *Fundamental Mathematical Concepts for Machine Learning in Science*,
https://doi.org/10.1007/978-3-031-56431-4

G

H

I